Low-Profile Natural and Metamaterial Antennas

Low-Profile Natural and Metamaterial Antennas

Analysis Methods and Applications

Hisamatsu Nakano
Hosei University, Koganei, Tokyo

 IEEE Antennas and Propagation Society, *Sponsor*

The IEEE Press Series on Electromagnetic Wave Theory
Andreas C. Cangellaris, *Series Editor*

Library of Congress Cataloging-in-Publication Data is available.

ISBN: 978-1-118-85979-7

Contents

Preface

$\mathbf{T}$his book is written for antenna engineers, researchers, graduate students, and advanced undergraduate students, who need to realize low-profile antennas that are required for modern communication systems, such as mobile and earth-satellite communications systems.

Low-profile antennas have been an ongoing topic of interest in the antenna field for over four decades. Their inherent characteristics of compactness, low aerodynamic resistance, and durability make them attractive for use in advanced/smart base station antennas, vehicular antennas, satellite communication antennas, radar antennas, and the like. As these systems advance, the need for smaller, more robust low-profile designs to meet ever-stricter system requirements has driven the field of low-profile antenna design forward. With this trend has come the need for new antenna design references; this book is intended to help meet this need.

Normally, electromagnetic properties in nature are right-handed. Antennas with right-handed properties are designated as natural antennas. On the other hand, antennas having electromagnetic properties that are not found in naturally occurring materials are designated as metamaterial-based antennas (simply referred to as metamaterial antennas). Note that metamaterials are realized artificially, typically by arraying small conducting elements periodically in one, two, or three dimensions. These artificial materials are generally referred to as left-handed materials, because they obey the left-hand rule, as opposed to the right-hand rule observed for natural materials. This unique property leads to antenna implementations that otherwise could not be realized, including a number of low-profile antenna implementations.

Designing a low-profile antenna brings with it a unique challenge—overcoming the inherent tendency for antenna performance to decrease as the height is decreased. This book covers recent progress in low-profile natural and metamaterial antennas and is composed of three parts: Introduction (Part I), Low-Profile Natural Antennas (Part II), and Low-Profile Metamaterial-Based Antennas (Part III).

Part I has three chapters (Chapters 1–3). Chapter 1 defines the natural and metamaterial-based antennas to be covered in this book, based on the *propagation phase constant* of the current along the radiation element. Chapters 2 and 3 provide the fundamentals for readers to be able to write computer programs based on the method of moments (MoM) (Chapter 2) and the finite-difference time-domain method (FDTDM) (Chapter 3). For this, in Chapter 2, a series of integral equations (N-series integral equations of N1–N5) for the MoM are discussed. It is noted that an arbitrarily shaped conductor (of one, two, or three dimensions) is modeled as an aggregate of conducting wire cells (subdivided elements), and the periphery of each cell is approximated by straight wires or curved wires, to which the N-series integral equations are applied. In Chapter 3, the basic FDTDM is summarized and the locally one-dimensional (LOD) FDTDM is presented.

Part II is composed of 15 chapters (Chapters 4–18) and discusses low-profile natural antennas that are classified into four groups: base station antennas (Part II-1: Chapters 4–7), card antennas for mobile equipment (Part II-2: Chapters 8–10), beam-forming antennas (Part II-3: Chapters 11–15), and earth–satellite and satellite–satellite communications antennas (Part II-4: Chapters 16–18).

The first part of low-profile natural antennas covers base station antennas (II-1), where wideband operation for inverted-F antennas and multiband operation for multiloop antennas are discussed. In addition, the realization of ultra-wideband (UWB) operation for a fan-shaped antenna and of a body-of-revolution antenna with a shorted parasitic ring (BOR–SPR) is described.

The second part (II-2) presents topics on low-profile card antennas, including the realization of multiband operation for inverted-LFL card antennas, together with the realization of UWB operation for fan-shaped card antennas and planar monopole card antennas. These card antennas are designed to fit into the limited space in mobile equipment, such as portable telephones and personal computers.

The beamforming antenna discussion in Part II-3 starts with the realization of a reconfigurable antenna, where inverted-F elements above an electromagnetic band-gap reflector are used. Next, reconfigurability for a bent two-leaf (BeToL) antenna and a bent four-leaf (BeFoL) antenna is discussed. Using switching circuits, these antennas can be reconfigured to radiate a beam in one of several directions at a specific frequency, while maintaining the same radiation characteristics. It is empha-sized that the horizontal area of the BeToL and BeFoL antennas is much smaller than that of corresponding reconfigurable antennas that use patches.

The discussion is extended to beamforming based on the Fabry–Pérot principle. It is shown that a single low-gain feed patch forms a high-gain tilted LP beam in a specific direction, by placing a parasitic layer consisting of loop elements above the feed patch. This antenna structure is simple and differs from conventional tilted-beam array antenna structures, where the arrayed radiation elements are connected to a main feed source by transmission lines through phase shifters and attenuators to form the tilted beam.

The discussion in II-3 finishes with two grid array antennas: one is a linearly polarized (LP) rhombic grid array antenna and the other is a circularly polarized (CP) loop grid array antenna. These antennas have a radiation beam that scans from the broadside direction to the forward direction, with a high gain for both LP and CP waves.

In LP-wave communication systems, the receiving antenna must be aligned with the polarization direction of the transmitting antenna; for instance, if a ver-tically polarized antenna is used for a transmitting antenna, then a vertically polarized receiving antenna is needed to maximize the reception of the trans-mitted power. In other words, the transmitting and receiving antennas need to be aligned such that the polarization directions are the same. On the other hand, a communication system where a CP wave is used does not need such align-ment; for example, if a right-handed CP antenna is used as a transmitting antenna, then it is enough to point a right-handed CP antenna at the transmitting antenna.

It is from this point of view that the fourth part, II-4, which discusses earth–satellite and satellite–satellite communications antennas, treats low-profile CP antennas. Arrays of spiral, helical, and curl antennas that obtain very high aperture efficiency are presented. In addition, a low-profile composite spiral and helical antenna array, which forms a high-gain tilted beam that can be aimed toward a satellite, is discussed.

Part III is composed of five chapters (Chapters 19–23) and discusses low-profile metamaterial-based antennas. Chapter 19 reveals that a metamaterial-based straight-line antenna (metaline antenna) radiates an LP backward beam, which cannot be achieved with a corresponding natural straight-line antenna having a right-handed property. Chapter 20 presents a new finding that a metamaterial-based loop antenna (metaloop) radiates single- and dual-peak LP beams at frequencies below a specific frequency (transition frequency) in addition to radiating the same beams at frequencies above the transition frequency.

Subsequently, Chapter 21 presents an open metaloop antenna that radiates a left-handed CP beam across a specific frequency band and a right-handed CP beam across a different frequency band (dual-band counter CP radiation). It is noted that such dual-band counter CP radiation cannot be obtained using a corresponding natural loop antenna that has a fixed single feed point. This is also true for both *natural* spiral and *natural* helical antennas.

However, the metamaterial-based spiral (metaspiral) antenna presented in Chapter 22 and the metamaterial-based helical (metahelical) antenna presented in Chapter 23 are shown to create dual-band counter CP radiation, which solves this issue. It should be emphasized that the antenna height for the metaspiral antenna is approximately 1/100 of the wavelength at the lowest operating frequency, in contrast to the 1/4 wavelength antenna height of conventional antennas backed by a conducting plate (reflector).

It is hoped that the antennas presented in this book will give readers supplemental knowledge that will be useful for practical antenna applications.

HISAMATSU NAKANO

Acknowledgments

I would like to thank the IEEE Press Liaison Committee and Wiley for encouraging me to publish this book. Thanks are also extended to Professor J. Yamauchi (Hosei University), Professor J. Shibayama (Hosei University), and V. Shkawrytko (Studio Victor Co.) for their support in the preparation of this manuscript.

Special thanks are expressed to the members of Nakano laboratory, Hosei University, who were involved in the realization of this book: H. Mimaki (Chapters 1–23), S. Okabe (Chapter 1), T. Yoshida (Chapter 2), K. Sakata (Chapter 2), K. Anjo (Chapters 2, 3, and 9), J. Shibayama (Chapter 3), M.Toida (Chapters 4–6), M. Takeuchi (Chapter 7), H. Kawabe (Chapter 8), Y. Kobayashi (Chapters 8 and 20), Y. Oka (Chapter 9), K. Iyoku (Chapter 10), R. Takebe (Chapter 11), R. Okamura (Chapter 11), R. Kato (Chapter 11), Y. Iitsuka (Chapters 12 and 13), T. Kawano (Chapters 12 and 13), Y. Sato (Chapter 14 and 16), S. Mitsui (Chapter 15), H. Kataoka (Chapter 19), K. Sakata (Chapter 19), M. Miura (Chapter 20), K. Yoshida (Chapter 21), T. Shimizu (Chapter 22), Y. Okuyama (Chapter 22), K. Monma (Chapter 23), and M. Tanaka (Chapter 23).

Finally, very special thanks are expressed to Professor T. Uno (Tokyo University of Agriculture and Technology), Professor K. Noguchi (Kanazawa Institute of technology), Associate professor T. Arima (Tokyo University of Agriculture and Technology), and PhD graduate student K. Fujita (Chuo University) for their contribution to the exercise sections in this book.

HISAMATSU NAKANO

Part I

Introduction

Chapter 1

Categorization of Natural Materials and Metamaterials

An electromagnetic material is categorized by its constitutional parameters, permittivity ε and permeability μ. A double-positive (DPS) material ($\varepsilon > 0$ and $\mu > 0$) is defined as a right-handed (RH) material. The phase constant of wave propagation within the RH material exhibits a positive value ($\beta > 0$). A double-negative (DNG) material ($\varepsilon < 0$ and $\mu < 0$) is defined as a left-handed (LH) material. The phase constant of wave propagation within the LH material exhibits a negative value ($\beta < 0$). Note that β within a mu-negative (MNG) material ($\varepsilon > 0$ and $\mu < 0$) and an epsilon-negative (ENG) material ($\varepsilon < 0$ and $\mu > 0$) is zero (i.e., evanescent).

A DPS material is a material found easily in nature and called a natural material, while a DNG, MNG, or ENG material is an artificial material and called a metamaterial (MTM) [1].

1.1 NATURAL AND METAMATERIAL ANTENNAS DISCUSSED IN THIS BOOK

Most antennas are made of natural materials. Antennas based on metamaterials are new, and some examples are found in Refs [1–3]. The categorization of natural and metamaterial antennas presented *in this book* is in reference to β, the propagation phase constant of the *current* flowing along a *fed* element.

Figure 1.1a shows a fed antenna where the out-going current from the feed point F toward the antenna element ends flows with a positive phase constant ($\beta > 0$). This means that the phase distribution takes a regressive form, that is, the phase is delayed from point F toward the antenna element ends. This type of antenna is categorized as a *natural* (NTR) *antenna*.

Figure 1.1b shows a fed antenna where the propagation phase constant of the out-going current can be either negative within a specific frequency band ($\beta < 0$) or zero at a nonzero frequency ($\beta = 0$). This type of antenna is categorized as a *metamaterial-based antenna* (simply referred to as a metamaterial antenna). The phase

Low-Profile Natural and Metamaterial Antennas: Analysis Methods and Applications, First Edition.
Hisamatsu Nakano.

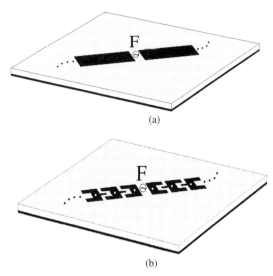

Figure 1.1   Antenna definition. (a) Natural antenna. The propagation phase constant is positive. (b) Metamaterial-based antenna (metamaterial antenna). The propagation phase constant is either negative within a specific frequency band or zero at a nonzero specific frequency.

distribution for $\beta < 0$ takes a progressive form from point F to the antenna element ends. A phase constant of zero ($\beta = 0$) means that the wavelength is infinitely long.

Exercise

Figure 1.2a shows a plane wave traveling within a lossless medium (permittivity c and permeability μ) in the z-direction. Figure 1.2b shows a lossless transmission line characterized by distributed circuit parameters [$C'(F/m),L'(H/m),C'_Z(F \bullet m)$, and $L'_Y(H \bullet m)$]. Discuss the correspondence between the medium constitutional parameters (ε and μ) and the circuit parameters.

Answer   Plane wave propagation within a lossless medium is specified by the following characteristic impedance Z_C and propagation constant γ:

$$Z_C = \sqrt{\frac{\mu}{\epsilon}} \tag{1.1}$$

$$\gamma = j\omega\sqrt{\mu\epsilon} \tag{1.2}$$

The propagation of voltage and current in a lossless transmission line is specified by the following characteristic impedance Z_{MTM} and propagation constant γ_{MTM}:

$$Z_{MTM} = \sqrt{\frac{Z'}{Y'}} \tag{1.3}$$

$$\gamma_{MTM} = \sqrt{Z' \, Y'} \tag{1.4}$$

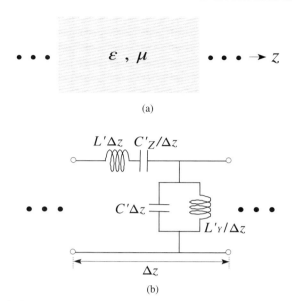

(a)

(b)

Figure 1.2   Equivalence. (a) Plane wave propagation within a lossless medium. (b) Lossless transmission line.

where

$$Z' = j\omega \left(L' - \frac{1}{\omega^2 C'_Z} \right) \equiv j\omega\,\mu_{TL} \qquad (1.5)$$

$$Y' = j\omega \left(C' - \frac{1}{\omega^2 L'_Y} \right) \equiv j\omega\,\epsilon_{TL} \qquad (1.6)$$

Then, Eqs. (1.3) and (1.4) are given by

$$Z_{MTM} = \sqrt{\frac{\mu_{TL}}{\epsilon_{TL}}} \qquad (1.7)$$

$$\gamma_{MTM} = j\omega\sqrt{\mu_{TL}\epsilon_{TL}} \qquad (1.8)$$

It is concluded that the medium constitutional parameters μ and ϵ correspond to the circuit parameters μ_{TL} and ϵ_{TL}, respectively,

$$\mu = \mu_{TL} = L' - \frac{1}{\omega^2 C'_Z} \qquad (1.9)$$

$$\epsilon = \epsilon_{TL} = C' - \frac{1}{\omega^2 L'_Y} \qquad (1.10)$$

Note that μ_{TL} and ϵ_{TL} can both be negative across a specific frequency region. In such a situation (simultaneously $\mu_{TL} < 0$ and $\epsilon_{TL} < 0$), the transmission line in Fig. 1.2b has a negative phase constant of $\beta = -\omega\sqrt{|\epsilon_{TL}||\mu_{TL}|}$. ■

1.2   SOME ANTENNA EXAMPLES

The above-mentioned categorization is explained using some examples. Figure 1.3 shows spiral antennas, a spiral with a cavity (Fig. 1.3a) [4], and a spiral antenna above an electromagnetic band gap (EBG) reflector (Fig. 1.3b) [5]. These antennas are fed from their center terminals. The phase constant for the out-going current along each antenna arm is positive $(\beta > 0)$. Hence, these antennas are categorized as natural antennas.

On the other hand, the current along the spiral arms in Fig. 1.3c shows a negative phase constant $(\beta < 0)$ across a specific frequency band, as will be discussed in Chapter 22 [6]. Hence, this antenna is categorized as a metamaterial antenna. The antenna shown in Fig. 1.3d is designed using a zero phase constant $(\beta = 0)$ at a nonzero frequency [7], and hence it is categorized as a metamaterial antenna.

Note that this book focuses on NTR and MTM antennas, as categorized above. Readers can find discussion on other MTM-related antennas [8–21], including an MTM-inspired antenna system composed of a fed element and an ENG or MNG metamaterial. The fed antennas used for the MTM-inspired antenna systems in Fig. 1.4a and b are, respectively, a natural monopole antenna and a natural patch antenna [the current of each antenna has a positive phase constant $(\beta > 0)$]. The effects of each MTM on the antenna system performance are discussed in Refs 20 and 21.

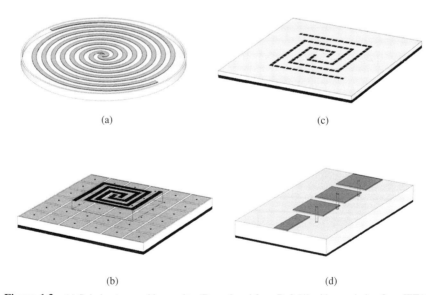

(a)   (c)

(b)   (d)

Figure 1.3   (a) Spiral antenna with a cavity. (Reproduced from Ref. [4] with permission from IET.) (b) Spiral antenna above an EBG reflector. (Reproduced from Ref. [5] with permission from IEEE.) (c) Metamaterial spiral (Metaspiral) antenna. (Reproduced from Ref. [6] with permission from IEEE.) (d) Zeroth-order resonance antenna. (Reproduced from Ref. [7] with permission from IEEE.)

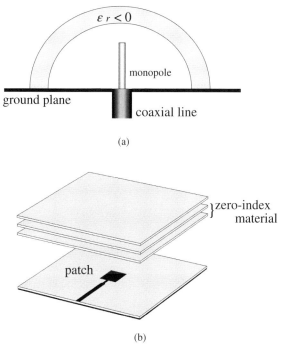

(a)

(b)

Figure 1.4   MTM-inspired antenna systems. (a) Monopole surrounded by an ENG shell. (Reproduced from Ref. [20] with permission from IEEE.) (b) Patch with MTM layers. (Reproduced from Ref. [21] with permission from IEEE.)

Exercise

Figure 1.5 shows a situation where a plane wave illuminates a slab of thickness B, effective relative permittivity ε_r, and effective relative permeability μ_r. The phase constant within the slab is given by $\beta = k_0 \sqrt{\mu_r \varepsilon_r}$, where $k_0 = \omega/c$ is the phase constant (real number) in free space, with ω and c being the angular frequency and the velocity of light, respectively. Express the wave impedance within the slab, using the scattering parameters S_{11} and S_{21} [22].

Answer   Using a signal flow chart [23], the scattering parameters for a finite thickness slab are written as

$$S_{11} = \frac{(1 - \varsigma^2)\Gamma}{1 - \Gamma^2 \varsigma^2} \tag{1.11}$$

$$S_{21} = \frac{(1 + \Gamma)(1 - \Gamma)\varsigma}{1 - \Gamma^2 \varsigma^2} = \frac{(1 - \Gamma^2)\varsigma}{1 - \Gamma^2 \varsigma^2} \tag{1.12}$$

where

$$\Gamma = \frac{z - 1}{z + 1} \tag{1.13}$$

$$\varsigma = e^{-j\omega \frac{1}{c}\sqrt{\mu_r \varepsilon_r}B} \tag{1.14}$$
$$= e^{-jk_0 nB}$$

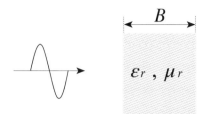

Figure 1.5 A plane wave illuminating a slab of effective relative permittivity ε_r, effective relative permeability μ_r, and thickness B.

Note that z in Eq. (1.13) is the wave impedance within the slab normalized to the free-space wave impedance Z_0, and n in Eq. (1.14) is the refractive index.

From Eqs. (1.11) and (1.12), Γ is expressed as

$$\Gamma = \xi \pm \sqrt{\xi^2 - 1}, \tag{1.15}$$

where

$$\xi = \frac{1 - (S_{21}^2 - S_{11}^2)}{2S_{11}} \tag{1.16}$$

The normalized wave impedance z in Eq. (1.13) is calculated using Γ from Eq. (1.15):

$$z = \pm\sqrt{\frac{(1 + S_{11})^2 - S_{21}^2}{(1 - S_{11})^2 - S_{21}^2}} \tag{1.17}$$

where the sign should be chosen so that the real part of z is

$$\text{Re}[z] \geq 0. \tag{1.18}$$

■

REFERENCES

1. N. Engahta and R. Ziolkowski, *Metamaterials*, NJ: Wiley, 2006.
2. C. Caloz and T. Itoh, *Electromagnetic metamaterials*, NJ: Wiley, 2006.
3. G. Eleftheriades and K. Balmain, *Negative-refraction metamaterials: fundamental principles and applications*, NJ: Wiely, 2005.
4. H. Nakano, S. Sasaki, H. Oyanagi, and J. Yamauchi, Cavity-backed Archimedean spiral antenna with strip absorber. *IET Microw. Antennas Propag.*, vol. 2, no. 7, pp. 725–730, 2008.
5. H. Nakano, K. Hitosugi, N. Tatsuzawa, D. Togashi, H. Mimaki, and J. Yamauchi, Effects on the radiation characteristics of using a corrugated reflector with a helical antenna and an electromagnetic band-gap reflector with a spiral antenna. *IEEE Trans. Antennas Propag.*, vol. 53, no. 1, pp. 191–199, 2005.
6. H. Nakano, J. Miyake, M. Oyama, and J. Yamauchi, Metamaterial spiral antenna. *IEEE Antennas Wirel. Propag. Lett.*, vol. 10, pp. 1555–1558, 2011.

7. A. Lai, M. K. H. Leong, and T. Itoh, Infinite wavelength resonant antennas with monopolar radiation pattern based on periodic structures. *IEEE Trans. Antennas Propag.*, vol. 55, no. 3, pp. 868–876, 2007.

8. M. A. Antoniades and G. V. Eleftheriades, A folded-monopole model for electrically small NRI-TL metamaterial antennas. *IEEE Antennas Wirel. Propag. Lett.*, vol. 7, pp. 425–428, 2008.

9. M. A. Antoniades and G. V. Eleftheriades, A broadband dual-mode monopole antenna using NRI-TL metamaterial loading. *IEEE Antennas Wirel. Propag. Lett.*, vol. 8, pp. 258–261, 2009.

10. F. Qureshi, M. A. Antoniades, and G. V. Eleftheriades, Compact and low-profile metamaterial ring antenna with vertical polarization. *IEEE Antennas Wirel. Propag. Lett.*, vol. 4, pp. 333–336, 2005.

11. J. Zhu, M. A. Antoniades, and G. V. Eleftheriades, A compact tri-band monopole antenna with single-cell metamaterial loading, *IEEE Trans. Antennas Propag.*, vol. 58, no. 4, pp. 1031–1038, 2010.

12. J.-H. Park, Y.-H. Ryu, J. G. Lee, and J. H. Lee, Epsilon negative zeroth-order resonator antenna. *IEEE Trans. Antennas Propag.*, vol. 55, pp. 3710–3712, 2007.

13. W. Liu, Z. N. Chen, and X. Qing, Metamaterial-based low-profile broadband mushroom antenna. *IEEE Trans. Antennas Propag.*, vol. 62, no. 3, pp. 1165–1172, 2014.

14. Nasimuddin, Z. N. Chen, and X. Qing, Substrate integrated metamaterial-based leak-wave antenna with improved boresight radiation bandwidth. *IEEE Trans. Antennas Propag.*, vol. 61, no. 7, pp. 3451–3457, 2013.

15. M. Sun, Z. N. Chen, and X. Qing, Gain enhancement of 60-GHz antipodal tapered slot antenna using zero-index metamaterial. *IEEE Trans. Antennas Propag.*, vol. 61, no. 4, pp. 1741–1746, 2013.

16. Nasimuddin, Z. N. Chen, and X. Qing, Multilayered composite right/left-handed leaky-wave antenna with consistent gain. *IEEE Trans. Antennas Propag.*, vol. 60, no. 11, pp. 5056–5062, 2012.

17. P. Jin and R. W. Ziolkowski, Metamaterial-inspired, electrically small, Huygens sources. *IEEE Antennas Wireless Propag. Lett.*, vol. 9, pp. 501–505, 2010b.

18. P. Jin and R. W. Ziolkowski, Multi-frequency, linear and circular polarized, metamaterial-inspired near-field resonant parasitic antennas. *IEEE Trans. Antennas Propag.*, vol. 59, pp. 1446–1459, 2011.

19. N. Zhu and R. W. Ziolkowski, Active metamaterial-inspired broad bandwidth, efficient, electrically small antennas. *IEEE Antennas Wirel. Propag. Lett.*, vol. 10, pp. 1582–1585, 2011.

20. R. W. Ziolkowski and A. Erentok, Metamaterial-based efficient electrically small antennas. *IEEE Trans. Antennas Propag.*, vol. 54, no. 7, pp. 2113–2130, 2006.

21. D. Li, Z. Szabó, X. Qing, Er.-P. Li, and Z. N. Chen, A high gain antenna with an optimized metamaterial inspired superstrate. *IEEE Trans. Antennas Propag.*, vol. 60, no. 12, pp. 6018–6023, 2012.

22. A. M. Nicolson and G. F. Ross, Measurement of the intrinsic properties of materials by time-domain techniques. *IEEE Trans. Instrum. Meas.*, vol. IM-19, no. 4, pp. 377–382, 1970.

23. D. Pozar, *Microwave engineering*, second edition, Wily, NY, 1998.

Chapter 2

Integral Equations and Method of Moments

2.1 BASIC ANTENNA CHARACTERISTICS

The electric field $\mathbf{E}$ radiated from an antenna (volume current density $\mathbf{J}$) located in free space is calculated using the vector potential $\mathbf{A}$ [5]:

$$\mathbf{E} = \frac{1}{j\omega\mu_0\varepsilon_0}\left[\nabla\nabla\bullet\mathbf{A} + k^2\mathbf{A}\right] \tag{2.1}$$

where ω is the angular frequency; μ_0 and ε_0 are the permeability and permittivity in free space, respectively; and k is $\omega\sqrt{\mu_0\varepsilon_0}$. Let $\mathbf{r}$ and $\mathbf{r'}$ be vectors locating observation and source points, respectively. When $|\mathbf{r}|\ (\equiv r)$ is much greater than $|\mathbf{r'}|$ and $k|\mathbf{r}| = kr \gg 1$, Eq. (2.1) becomes

$$\mathbf{E} = -\frac{j\omega\mu_0 e^{-jkr}}{4\pi r}\int_V\left[\mathbf{J}(\mathbf{r'}) - \{\hat{\mathbf{r}}\bullet\mathbf{J}(\mathbf{r'})\}\hat{\mathbf{r}}\right]e^{jk\hat{\mathbf{r}}\bullet\mathbf{r'}}\,dV' \tag{2.2}$$

where $\hat{r}$ is a unit vector in the radial direction: $\hat{\mathbf{r}} \equiv \mathbf{r}/|\mathbf{r}|$. The field $\mathbf{E}$ in Eq. (2.2) is called the *far-zone field* or *radiation field* and has θ and ϕ components, E_θ and E_ϕ, in the spherical coordinate system (r, θ, ϕ). The radial component, E_r, does not exist.

The tip of the far-zone electric field vector at a given position varies with time. In general, the locus of this tip motion describes an ellipse, which is specified by the major axis $2E_{max}$ and minor axis $2E_{min}$. The ratio of E_{max}/E_{min} is called the *axial ratio*, AR. The AR is expressed as

$$AR = 10\log_{10}\frac{Lax}{Sax}\,[\text{dB}] \tag{2.3}$$

where $Lax = |E_\phi|^2\sin^2\tau + |E_\theta|^2\cos^2\tau + |E_\phi||E_\theta|\cos\delta\sin2\tau$ and $Sax = |E_\phi|^2\cos^2\tau + |E_\theta|^2\sin^2\tau - |E_\phi||E_\theta|\cos\delta\sin2\tau$. Note that δ is the phase difference between E_θ and

Low-Profile Natural and Metamaterial Antennas: Analysis Methods and Applications, First Edition.
Hisamatsu Nakano.
© 2016 The Institute of Electrical and Electronics Engineers, Inc. Published 2016 by John Wiley & Sons, Inc.

E_ϕ, and τ is defined as

$$2\tau = \tan^{-1} \frac{2|E_\phi||E_\theta|\cos\delta}{|E_\theta|^2 - |E_\phi|^2} \tag{2.4}$$

The ratio of the maximum radiation intensity U_{max} (watts/m^2) to the average radiation intensity U_{av} is defined as the *directivity*, D. The directivity is written as $D = U_{max}/U_{av} = U_{max}/(P_{rad}/4\pi)$, where P_{rad} is the power radiated from the antenna. Note that the directivity does not convey information about the radiation efficiency, $\eta_{rad} = P_{rad}/P_{in}$, where P_{in} is the total power input to the antenna.

The (maximum) gain relative to an isotropic source is defined as $G = U_{max}/U_{iso} = U_{max}/(P_{in}/4\pi) = \eta_{rad}D$, where U_{iso} is the radiation intensity of a lossless isotropic source with the same power input P_{in}. The gain in terms of the far-zone electric field, $\mathbf{E}(r, \theta, \phi)$, is expressed as

$$G = \eta_{rad} \frac{4\pi|\mathbf{E}_{max}|^2}{\int_0^{2\pi}\int_0^\pi |\mathbf{E}(r,\theta,\phi)|^2 \sin\theta \, d\theta \, d\phi} \tag{2.5}$$

The *input impedance* of an antenna, $Z_{in} = R_{in} + jX_{in}$, is defined as the ratio of the voltage to the current at a pair of terminals. The real part of the input impedance is expressed as $R_{in} = R_{rad} + R_{loss}$, where R_{rad} is the radiation resistance describing the radiation from the antenna and R_{loss} is the loss resistance describing the ohmic losses on the antenna structure. The imaginary part, X_{in}, is related to the reactive near-field power. The input impedance should be designed to match the characteristic impedance of the feed line for maximum power transmission.

Note that the antenna characteristics, including the radiation field, axial ratio, gain, and input impedance, are all evaluated on the basis of the current distribution. Therefore, the accuracy of the antenna characteristics strongly depends on the determination of the correct current distribution. Integral equations to obtain the current distribution using the method of moments (MoM) are presented in Sections 2.2–2.9.

Exercise [24]

An electrically small spherical antenna (ESA) of radius R is located at the origin of a spherical coordinate system. This antenna consists of homogeneous material represented by the relative complex permittivity $\varepsilon_r(= \varepsilon_r' - j\varepsilon_r'')$. The field equations for the electric and magnetic fields (**E** and **H**) outside the ESA are expanded using vector spherical waves as [1]

$$\mathbf{E} = Z_0 \sum_{n=1}^\infty \sum_{m=-n}^n \frac{1}{\sqrt{n(n+1)}} \left(\frac{-j}{k} a_{nm} \nabla \times (h_n^{(2)}(kr)\mathbf{L}Y_n^m) + b_{nm}h_n^{(2)}(kr)\mathbf{L}Y_n^m \right) \tag{2.6}$$

$$\mathbf{H} = \sum_{n=1}^\infty \sum_{m=-n}^n \frac{1}{\sqrt{n(n+1)}} \left(a_{nm}h_n^{(2)}(kr)\mathbf{L}Y_n^m + \frac{j}{k}b_{nm} \nabla \times (h_n^{(2)}(kr)\mathbf{L}Y_n^m) \right) \tag{2.7}$$

where $k(= \omega\sqrt{\varepsilon_0\mu_0})$ and $Z_0(= \sqrt{\mu_0/\varepsilon_0})$ are the wave number and characteristic impedance for free space, respectively; $h_n^{(2)}$ and Y_n^m are the spherical Hankel function of the second order and the spherical harmonics [2], respectively; and $\mathbf{L}$ is a differential operator defined as $\mathbf{L} = j\mathbf{r} \times \nabla$ with $\mathbf{r} = r\hat{\mathbf{r}}$ ($\hat{\mathbf{r}}$ is a unit vector). Coefficients a_{nm} and b_{nm} are written as [2]

$$a_{nm} = \int_v \mathbf{J} \cdot \mathbf{U}_{nm}^* dv \tag{2.8}$$

$$b_{nm} = \int_v \mathbf{J} \cdot \mathbf{V}_{nm}^* dv \tag{2.9}$$

where $\mathbf{J}$ is the antenna volume current density and

$$\mathbf{U}_{nm} = \frac{k}{\sqrt{n(n+1)}} \left(\frac{\partial}{\partial r}(rj_n(kr))\nabla Y_n^m + n(n+1)\frac{j_n(kr)}{r} Y_n^m \hat{\mathbf{r}} \right) \tag{2.10}$$

$$\mathbf{V}_{nm} = \frac{jk^2}{\sqrt{n(n+1)}} j_n(kr)\nabla Y_n^m \times \mathbf{r} \tag{2.11}$$

Note that j_n is the spherical Bessel function, and the norms of $\mathbf{U}_{nm}$ and $\mathbf{V}_{nm}$ are [3]

$$\|\mathbf{U}_{nm}\|^2 = \int_v \mathbf{U}_{nm} \bullet \mathbf{U}_{nm}^* dv = k \left\{ \frac{(kR)^3}{2} \left[j_{n-1}^2(kR) - j_n(kR)j_{n-2}(kR) \right] - n(kR)j_n^2(kR) \right\} \tag{2.12}$$

$$\|\mathbf{V}_{nm}\|^2 = \int_v \mathbf{V}_{nm} \bullet \mathbf{V}_{nm}^* dv = \frac{k}{2}(kR)^3 \left[j_n^2(kR) - j_{n+1}(kR)j_{n-1}(kR) \right] \tag{2.13}$$

The radiated power of the antenna, P_{rad}, is obtained using Eqs. (2.6) and (2.7):

$$P_{rad} = \frac{1}{2} \mathrm{Re} \left[\int_S \mathbf{E} \times \mathbf{H}^* \bullet d\mathbf{S} \right]$$
$$= \frac{Z_0}{2k^2} \sum_{n=1}^{\infty} \sum_{m=-n}^{+n} \left(|a_{nm}|^2 + |b_{nm}|^2 \right) \tag{2.14}$$

The material loss P_{loss} is given as

$$P_{loss} = \frac{1}{2} \mathrm{Re} \left[\int_v \mathbf{E} \bullet \mathbf{J}^* dv \right]$$
$$= \frac{1}{2} \int_v \mathrm{Re} \left[\frac{1}{j\omega(\varepsilon_0 \varepsilon_r - \varepsilon_0)} \right] |\mathbf{J}|^2 dv \tag{2.15}$$
$$= \frac{\pi Z_0}{k} \frac{1}{M} \int_V |\mathbf{J}|^2 dv$$

where

$$M = \frac{2\pi |\varepsilon_r - 1|^2}{\varepsilon_r''} \tag{2.16}$$

Note that $\mathbf{E} = \mathbf{J}/j\omega\varepsilon_0(\varepsilon_r - 1)$ is used for Eq. (2.15), where the conductivity of the material is included in ε_r''. Derive the maximum radiation efficiency of this antenna.

Answer   The volume current density can be written as

$$\mathbf{J} = \sum_{n=1}^{\infty} \sum_{m=-n}^{+n} \left[\frac{a_{nm}}{\|\mathbf{U}_{nm}\|^2} \mathbf{U}_{nm} + \frac{b_{nm}}{\|\mathbf{V}_{nm}\|^2} \mathbf{V}_{nm} \right] + \mathbf{J}_{NR} \tag{2.17}$$

where $\mathbf{J}_{NR}$ is the nonradiating current component, holding $\int_v \mathbf{J}_{NR} \cdot \mathbf{U}_{nm}^* dv = \int_v \mathbf{J}_{NR} \cdot \mathbf{V}_{nm}^* dv = 0$. Substituting Eq. (2.17) into Eq. (2.15), the material loss is

$$P_{loss} = \frac{\pi Z_0}{k} \frac{1}{M} \left[\sum_{n=1}^{\infty} \sum_{m=-n}^{+n} \left(\frac{|a_{nm}|^2}{\|\mathbf{U}_{nm}\|^2} + \frac{|b_{nm}|^2}{\|\mathbf{V}_{nm}\|^2} \right) + \int_V |\mathbf{J}_{NR}|^2 dv \right] \tag{2.18}$$

The definition of the radiation efficiency is

$$\eta_{rad} = \frac{P_{rad}}{P_{rad} + P_{loss}} \tag{2.19}$$

When $\mathbf{J}_{NR} = 0$ (ideal case), Eq. (2.19) is expressed as

$$\eta_{rad} = \left(1 + \frac{2\pi}{M} \hat{D} \right)^{-1} \tag{2.20}$$

where

$$\hat{D} = \frac{\displaystyle\sum_{n=1}^{\infty} \sum_{m=-n}^{n} \left(\frac{|a_{nm}|^2}{\|\mathbf{U}_{nm}\|^2/k} + \frac{|b_{nm}|^2}{\|\mathbf{V}_{nm}\|^2/k} \right)}{\displaystyle\sum_{n=1}^{\infty} \sum_{m=-n}^{n} \left(|a_{nm}|^2 + |b_{nm}|^2 \right)} \tag{2.21}$$

It is noted that the radiation efficiency is maximal when $\hat{D}$ is minimal, which is obtained by determining

$$\max_{m,n}[\|\mathbf{U}_{nm}\|^2, \|\mathbf{V}_{nm}\|^2] \tag{2.22}$$

For this determination, the following relations are used.

$$\frac{\|\mathbf{U}_{nm}\|^2}{k} - \frac{\|\mathbf{U}_{(n+1)m}\|^2}{k}$$
$$= (n+1)kRj_{n+1}^2(kR) + kRj_n(kR)\left[j_n(kR) + kR\frac{d}{d(kR)}j_n(kR) \right] > 0, \quad (if\ kR \leq 2.7) \tag{2.23}$$

$$\frac{\|\mathbf{V}_{nm}\|^2}{k} - \frac{\|\mathbf{V}_{(n+1)m}\|^2}{k} = (kR)^2 j_n(kR)j_{n+1}(kR) > 0, \quad (if\ kR \leq 2.7) \tag{2.24}$$

$$\frac{\|\mathbf{U}_{nm}\|^2}{k} - \frac{\|\mathbf{V}_{nm}\|^2}{k} = kRj_n(kR)\left[j_n(kR) + kR\frac{d}{d(kR)}j_n(kR) \right] > 0, \quad (if\ kR \leq 2.7) \tag{2.25}$$

Based on Eqs. (2.23)–(2.25), Eq. (2.22) becomes

$$\max_{m,n} \left[\|\mathbf{U}_{nm}\|^2, \|\mathbf{V}_{nm}\|^2 \right] = \|\mathbf{U}_{1m}\|^2 \tag{2.26}$$

Therefore, the minimum $\hat{D}$ is obtained when

$$a_{nm} \begin{cases} \neq 0 & for\ n = 1\ and\ m \in \{-1, 0, +1\} \\ = 0 & for\ \forall n \in \{2, 3, \dots\}\ and\ \forall m \in \{0, \pm1, \pm2, \dots\} \end{cases}$$
$$b_{nm} = 0 \quad for\ \forall n \in \{1, 2, 3, \dots\}\ and\ \forall m \in \{0, \pm1, \pm2, \dots\} \tag{2.27}$$

Based on Eq. (2.27), the maximum radiation efficiency is derived:

$$\eta_{rad:max} = \left[1 + \frac{2\pi}{M}\left(\|\mathbf{U}_{1m}\|^2/k\right)^{-1}\right]^{-1} \tag{2.28}$$

where $\|\mathbf{U}_{10}\|^2/k = \|\mathbf{U}_{1(+1)}\|^2/k = \|\mathbf{U}_{1(-1)}\|^2/k$ is used. ∎

2.2 INTEGRAL EQUATION ON A STRAIGHT-WIRE ANTENNA

Figure 2.1 shows a straight-wire antenna located in a medium of permittivity ε and permeability μ. The radius and total length of the wire are a and $2L$, respectively. Three assumptions are made here: (1) the wire is perfectly conducting, (2) the radius a is small compared with the free-space wavelength λ ($a \ll \lambda$), and (3) the total length is large compared with a ($a \ll 2L$). Assumptions (2) and (3) are often called the *thin-wire approximation*, which ensures that the antenna characteristics can be calculated simply from the current flowing in the antenna axis direction [5].

The electric field $\mathbf{E}$ emanating from the current flowing along the wire is given by

$$\mathbf{E} = -grad\,\phi - j\omega\mathbf{A} \tag{2.29}$$

where ϕ and $\mathbf{A}$ are scalar and vector potentials, respectively, and $\omega = 2\pi f$ ($f =$ frequency). The z-axis component in Eq. (2.29) is written as

$$E_z = \frac{1}{j\omega\varepsilon}\int_{-L}^{+L}\left[\frac{\partial^2 V(z,z')}{\partial z^2} + \beta^2 V(z,z')\right]I(z')dz' \tag{2.30}$$

where $\beta = \omega\sqrt{\mu\varepsilon}$ and $I(z')$ are the current (line distribution) at z'. V is defined as

$$V(z,z') = \frac{1}{4\pi}\frac{e^{-j\beta r(z,z')}}{r(z,z')} \tag{2.31}$$

where

$$r(z,z') = \sqrt{a^2 + (z-z')^2} \tag{2.32}$$

The boundary condition at the surface of the perfectly conducting wire is given by

$$E_z + E_z^i = 0 \tag{2.33}$$

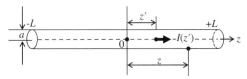

Figure 2.1   A straight-wire conductor.

where E_z^i is the z-axis component of the incident electric field or the applied electric field. Substituting Eq. (2.30) into Eq. (2.33), we have

$$\frac{1}{j\omega\varepsilon} \int_{-L}^{+L} I(z') \left[\frac{\partial^2 V(z, z')}{\partial z^2} + \beta^2 V(z, z') \right] dz' + E_z^i(z) = 0 \tag{2.34}$$

Eq. (2.34) is called *Pocklington's integral equation* [4–6]. Note that $V(z, z')$ and $E_z^i(z)$ in the integral equation are known functions, and $I(z')$ is an unknown function. In the following section, a method for determining the unknown $I(z')$ is discussed.

Exercise

Derive Eq. (2.30).

Answer   From Eq. (2.29), the z-component of the electric field, E_z, is

$$Ez = -j\omega A_z - \frac{\partial \phi}{\partial z} \tag{2.35}$$

where A_z is the z-component of the vector potential **A**. Applying the Lorentz condition

$$div\ \mathbf{A} + j\omega\mu\varepsilon\phi = 0 \tag{2.36}$$

to Eq. (2.35) leads to

$$Ez = \frac{1}{j\omega\varepsilon\mu} \left(\frac{\partial^2 A_z}{\partial z^2} + \beta^2 A_z \right) \tag{2.37}$$

where

$$A_z = \mu \int_{L_a} I(z') V(z, z') dz' \tag{2.38}$$

Note that the integral sign with L_a denotes integration over the antenna length.   ∎

2.3   METHOD OF MOMENTS

The current distribution $I(z')$ in Pocklington's integral equation (2.34) is obtained using the *method of moments* [5,7,8], which requires two steps. The first step in the MoM is to expand the unknown current $I(z')$:

$$I(z') = \sum_n I_n J_n(z') \tag{2.39}$$

where $J_n(z')(n = 1, 2, \ldots, N)$ are called *expansion functions*, and $I_n(n = 1, 2, \ldots, N)$ are called *expansion function coefficients*. Let the expansion functions be

$$J_n(z') = \begin{cases} J_n(z') & \text{for } z' \text{ in region } \Delta z_n \\ 0 & \text{otherwise} \end{cases} \tag{2.40}$$

By substituting Eq. (2.39) into the left side of Eq. (2.34) under the conditions of Eq. (2.40), the residual $R(z)$ is defined as

$$R(z) = \sum_n I_n \int_{\Delta z_n} J_n(z')\Pi(z, z')dz' + E_z^i(z) \tag{2.41}$$

where

$$\Pi(z, z') = \frac{1}{j\omega\varepsilon}\left[\frac{\partial^2 V(z, z')}{\partial z^2} + \beta^2 V(z, z')\right] \tag{2.42}$$

The second step in the MoM is to force the integration of each weighted residual to be zero:

$$\int_{L_a} W_m(z)R(z)dz = 0 \qquad (m = 1, 2, \ldots, N) \tag{2.43}$$

where $W_m(z)(m = 1, 2, \ldots, N)$ are called *weighting functions* or *testing functions*. Let the weighting functions be

$$W_m(z) = \begin{cases} W_m(z) & \text{for } z \in \Delta z_m \\ 0 & \text{otherwise} \end{cases} \tag{2.44}$$

Then, Eq. (2.43) can be expressed as

$$\sum_n I_n \int_{\Delta Z_m} W_m(z) \int_{\Delta Z_n} J_n(z')\Pi(z, z')dz'dz = -\int_{\Delta Z_m} W_m(z)E_z^i(z)dz \tag{2.45}$$

This equation is a simultaneous equation for $I_n(n = 1, 2,\ldots, N)$, and can be written as

$$\sum_n Z_{m,n}I_n = V_m(m = 1, 2, \ldots, N) \tag{2.46}$$

where

$$Z_{m,n} = \int_{\Delta Z_m} W_m(z) \int_{\Delta Z_n} J_n(z')\Pi(z, z')dz'dz \tag{2.47}$$

$$V_m = -\int_{\Delta Z_m} W_m(z)E_z^i(z)dz \tag{2.48}$$

The matrix form of Eq. (2.46) is expressed as

$$\left[Z_{m,n}\right][I_n] = [V_m] \tag{2.49}$$

where $\left[Z_{m,n}\right]$ and $[V_m]$ are called the *generalized impedance matrix* and the *generalized voltage matrix*, respectively. Thus, the expansion function coefficients $I_n(n = 1, 2, \ldots, N)$ can be written as

$$[I_n] = \left[Z_{m,n}\right]^{-1}[V_m] \tag{2.50}$$

Substituting these I_n into Eq. (2.39) yields the current distribution $I(z')$.

Exercise

When the same functions are used for both the expansion and weighting functions, the method of moments is often referred to as *Galerkin's method*. The expansion functions must be chosen such that they are close to the real current distribution, in order to get fast convergence in the numerical solution. Figure 2.2 illustrates the case where the expansion functions are piecewise sinusoidal functions, which are frequently used in antenna analysis:

$$J_n(z') = \begin{cases} \dfrac{\sin\beta(z' - z_{n-1})}{\sin\beta(z_n - z_{n-1})} & \text{for } z_{n-1} \le z' \le z_n \\[2mm] \dfrac{\sin\beta(z_{n+1} - z')}{\sin\beta(z_{n+1} - z_n)} & \text{for } z_n \le z' \le z_{n+1} \\[2mm] 0 & \text{otherwise} \end{cases} \tag{2.51}$$

Reduce the double integral form for the impedance matrix element $Z_{m,n}$ in Eq. (2.47) into a single integral form using Galerkin's method. In addition, calculate V_m in Eq. (2.48) for a delta-gap source of voltage V_0 existing at $z = z_m$ [5].

Answer

$$Z_{m,n} = \int_{z_{m-1}}^{z_m} \frac{\sin\beta(z - z_{m-1})}{\sin\beta(z_m - z_{m-1})} E_n(z)\mathrm{d}z + \int_{z_m}^{z_{m+1}} \frac{\sin\beta(z_{m+1} - z)}{\sin\beta(z_{m+1} - z_m)} E_n(z)\mathrm{d}z \tag{2.52}$$

where

$$\begin{aligned} E_n(z) &= \int_{z_{n-1}}^{z_{n+1}} J_n(z')\prod(z, z')\mathrm{d}z' \\[2mm] &= \frac{j30}{\sin\beta(z_n - z_{n-1})}\left[\frac{e^{-j\beta R_n^-}}{R_n^-}\cos\beta(z_n - z_{n-1}) - \frac{e^{-j\beta R_{n-1}^-}}{R_{n-1}^-}\right] \\[2mm] &\quad + \frac{j30}{\sin\beta(z_{n+1} - z_n)}\left[\frac{e^{-j\beta R_n^-}}{R_n^-}\cos\beta(z_{n+1} - z_n) - \frac{e^{-j\beta R_n^+}}{R_n^+}\right] \end{aligned} \tag{2.53}$$

Note that R_{n-1}^-, R_n^-, and R_n^+ are the distances between the observation point and source points z_{n-1}, z_n, and z_{n+1}, respectively, as shown in Fig. 2.3:

$$\left(R_{n-1}^-\right)^2 = a^2 + (z - z_{n-1})^2 \tag{2.54}$$

$$\left(R_n^-\right)^2 = a^2 + (z - z_n)^2 \tag{2.55}$$

$$\left(R_n^+\right)^2 = a^2 + (z - z_{n+1})^2 \tag{2.56}$$

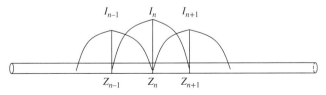

Figure 2.2   Piecewise sinusoidal functions.

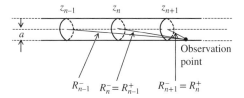

Figure 2.3   Distances between the observation point z and the source points z_{n-1}, z_n, and z_{n+1}.

The voltage matrix element V_m in Eq. (2.48) is

$$
\begin{aligned}
V_m &= -\int_{\Delta z_m} W_m(z) E_Z^i(z)\mathrm{d}z \\
&= -\int_{Z_{m-1}}^{Z_m} \frac{\sin\beta(z - z_{m-1})}{\sin\beta(z_m - z_{m-1})} V_0 \delta(z - z_m)\mathrm{d}z \\
&\quad -\int_{Z_m}^{Z_{m+1}} \frac{\sin\beta(z_{m+1} - z)}{\sin\beta(z_{m+1} - z_m)} V_0 \delta(z - z_m)\mathrm{d}z \\
&= -V_0
\end{aligned}
\tag{2.57}
$$

■

2.4   INTEGRAL EQUATION FOR AN ARBITRARILY SHAPED WIRE ANTENNA IN FREE SPACE

The method for determining the current distribution along a straight conductor has been discussed in Section 2.3. In this section, an integral equation for an arbitrarily shaped wire antenna is derived and solved for the unknown current along the wire using the MoM [5].

Figure 2.4 shows an arbitrarily shaped wire antenna, where s and s' denote the s-coordinates for an observation point and a source point along the wire,

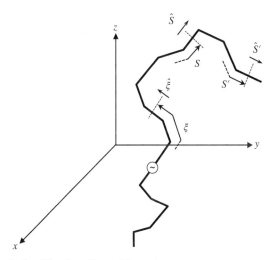

Figure 2.4   A conducting thin wire with an arbitrary shape.

respectively, and $\hat{s}$ and $\hat{s}'$ denote unit vectors tangential to the wire at s and s', respectively.

The s-component of the electric field generated by the current $I(s')$ on the wire is given by

$$E_s = -\frac{\partial \phi}{\partial s} - j\omega A_s \tag{2.58}$$

and the boundary condition at the surface of the conducting wire is written as

$$-j\omega\varepsilon E_s^i(s) = -j\omega\varepsilon \frac{\partial \phi(s)}{\partial s} + \beta^2 A_{s\mu}(s) \tag{2.59}$$

where $E_s^i(s)$ is the s-component of a field incident to the wire or the impressed electric field, and

$$A_{s\mu}(s) \equiv \frac{1}{\mu} A_s = \hat{s} \cdot \int_{L_a} I(s')\hat{s}' G(s,s')\mathrm{d}s' \tag{2.60}$$

$$\phi(s) = \frac{-1}{j\omega\varepsilon} \int_{L_a} \frac{\mathrm{d}I(s')}{\mathrm{d}s'} G(s,s')\mathrm{d}s' \tag{2.61}$$

$$G(s,s') = \frac{1}{4\pi} \frac{e^{-j\beta r(s,s')}}{r(s,s')} \tag{2.62}$$

Note that $r(s,s')$ in Eq. (2.62) is the distance between the observation point and the source point.

A scalar function Φ is defined as [9]

$$\Phi(s) \equiv -j\omega\varepsilon \int_0^s \phi(\xi)\mathrm{d}\xi \tag{2.63}$$

Then, the following equation is obtained:

$$\frac{\mathrm{d}^2\Phi(s)}{\mathrm{d}s^2} = -j\omega\varepsilon \frac{\mathrm{d}\Phi(s)}{\mathrm{d}s} \tag{2.64}$$

Substituting Eq. (2.64) into Eq. (2.59) results in

$$\frac{\mathrm{d}^2\Phi(s)}{\mathrm{d}s^2} = -\beta^2 A_{s\mu}(s) - j\omega\varepsilon E_s^i(s) \tag{2.65}$$

Adding $\beta^2\Phi(s)$ to both sides of Eq. (2.65) leads to

$$\frac{\mathrm{d}^2\Phi(s)}{\mathrm{d}s^2} + \beta^2\Phi(s) = \beta^2\left[\Phi(s) - A_{s\mu}(s)\right] - j\omega\varepsilon E_s^i(s) \tag{2.66}$$

The solution of Eq. (2.66) is

$$\begin{aligned}\Phi(s) = {}&C\cos\beta s + D\sin\beta|s| \\ &+ \int_0^s \beta\left[\Phi(\xi) - A_{\xi\mu}(\xi)\right]\sin\beta(s-\xi)\mathrm{d}\xi - \frac{j}{Z_0}\int_0^s E_\xi^i(\xi)\sin\beta(s-\xi)\mathrm{d}\xi\end{aligned} \tag{2.67}$$

where C and D are constants, and $Z_0 \equiv \sqrt{\mu/\varepsilon}$. Since $\Phi(0) = 0$ in Eq. (2.63), the constant C vanishes. After some manipulation, Eq. (2.67) becomes

$$
\int_{L_a} I(s')(\pi_1 - \pi_2 - \pi_3)\mathrm{d}s'
$$

$$
= D\sin\beta|s| - \frac{j}{Z_0}\int_0^s E_\xi^i(\xi)\sin\beta(s - \xi)\mathrm{d}\xi + \int_{L_a} I(s')G(0, s')\widehat{\mathbf{0}} \bullet \widehat{\mathbf{s}}'\cos\beta s\,\mathrm{d}s' \tag{2.68}
$$

where

$$
\pi_1 = G(s, s')\widehat{\mathbf{s}} \bullet \widehat{\mathbf{s}}' \tag{2.69}
$$

$$
\pi_2 = \int_0^s G(\xi, s')\frac{\mathrm{d}\widehat{\xi}}{\mathrm{d}\xi} \bullet \widehat{\mathbf{s}}'\cos\beta(s - \xi)\mathrm{d}\xi \tag{2.70}
$$

$$
\pi_3 = \int_0^s \left[\frac{\partial G(\xi, s')}{\partial\xi'}\left(\widehat{\xi} \bullet \widehat{\mathbf{s}}'\right) + \frac{\partial G(\xi, s')}{\partial s'}\right]\cos\beta(s - \xi)\mathrm{d}\xi \tag{2.71}
$$

The first term on the right side of Eq. (2.68) represents the effect of a slice genera-tor, and can be deleted when the integration of E_ξ^i is present. Therefore, Eq. (2.68) can be written as

$$
\int_{L_a} I(s')(\pi_1 - \pi_2 - \pi_3)\mathrm{d}s' = B\cos\beta s - \frac{j}{Z_0}\int_0^s E_\xi^i(\xi)\sin\beta(s - \xi)\mathrm{d}\xi \tag{2.72}
$$

where B is a constant to be determined by applying the condition that the current vanishes at the end of the antenna. Eq. (2.72) is called *Mei's integral equation* [5,9].

Exercise

Derive Eq. (2.72) from Eq. (2.67).

Answer   Manipulating the right side of Eq. (2.67) leads to

$$
\int_0^s \beta\Phi(\xi)\sin\beta(s - \xi)\mathrm{d}\xi = \Phi(s) + \int_{L_a}\int_0^s I(s')\frac{\partial G(\eta, s')}{\partial s'}\cos\beta(s - \eta)\mathrm{d}\eta\,\mathrm{d}s' \tag{2.73}
$$

and

$$
\int_0^s \beta A_{\xi\mu}(\xi)\sin\beta(s - \xi)\mathrm{d}\xi
$$

$$
= \int_{L_a} I(s')G(s, s')\widehat{\mathbf{s}} \bullet \widehat{\mathbf{s}}'\mathrm{d}s' - \int_{L_a} I(s')G(0, s')\widehat{\mathbf{0}} \bullet \widehat{\mathbf{s}}'\cos\beta s\,\mathrm{d}s'
$$

$$
- \int_0^s\int_{L_a}\left[\frac{\partial G(\xi, s')}{\partial\xi}\widehat{\xi} \bullet \widehat{\mathbf{s}}' + G(\xi, s')\frac{\partial\left(\widehat{\xi} \bullet \widehat{\mathbf{s}}'\right)}{\partial\xi}\right]I(s')\cos\beta(s - \xi)\mathrm{d}\xi\,\mathrm{d}s' \tag{2.74}
$$

Substituting Eq. (2.73) and Eq. (2.74) into Eq. (2.67) leads to

$$\int_{L_a} I(s')(\pi_1 - \pi_2 - \pi_3)\mathrm{d}s'$$

$$= D\sin\beta|s| - \frac{j}{Z_0}\int_0^s E_\xi^i(\xi)\sin\beta(s-\xi)\mathrm{d}\xi + \int_{L_a} I(s')G(0,s')\widehat{\mathbf{0}}\cdot\widehat{\mathbf{s}}'\cos\beta s\mathrm{d}s' \tag{2.75}$$

where

$$\pi_1 = G(s,s')\widehat{\mathbf{s}}\cdot\widehat{\mathbf{s}}' \tag{2.76}$$

$$\pi_2 = \int_0^s G(\xi,s')\frac{\widehat{\mathrm{d}\xi}}{\mathrm{d}\xi}\cdot\widehat{\mathbf{s}}'\cos\beta(s-\xi)\mathrm{d}\xi \tag{2.77}$$

$$\pi_3 = \int_0^s \left[\frac{\partial G(\xi,s')}{\partial\xi'}\left(\widehat{\boldsymbol{\xi}}\cdot\widehat{\mathbf{s}}'\right) + \frac{\partial G(\xi,s')}{\partial s'}\right]\cos\beta(s-\xi)\mathrm{d}\xi \tag{2.78}$$

The first term on the right side of Eq. (2.75) represents the effect of a slice generator, and can be deleted when the integration of E_ξ^i is present. Thus, Mei's integral equation (2.72) is obtained [9]. Note that Eq. (2.72) for a straight-wire antenna is reduced to

$$\int_{L_a} I(s')G(s,s')\mathrm{d}s' = B\cos\beta s - \frac{jV_0}{2Z_0}\sin\beta|s| \tag{2.79}$$

due to $\pi_2 = 0$ and $\pi_3 = 0$. Eq. (2.79) is called *Hallen's integral equation* [5,6,10].   ■

2.5   POINT-MATCHING TECHNIQUE

The unknown current $I(s')$ in integral equation (2.72) is often solved using a point-matching technique, which is an MoM technique. The point-matching technique starts with expanding the current $I(s')$ as

$$I(s') = \sum_n I_n J_n(s') \tag{2.80}$$

Substituting this into Eq. (2.72) leads to

$$\sum_n I_n \int_{L_a} J_n(s')\pi(s,s')\mathrm{d}s' = f(s) \tag{2.81}$$

where $\pi(s,s') \equiv \pi_1 - \pi_2 - \pi_3$, and $f(s)$ represents the right side of Eq. (2.72). Next, N different points $s_m(m = 1, 2, \ldots, N)$ along the antenna arm are selected and Eq. (2.81) is calculated at these points:

$$\sum_n I_n \int_{L_a} J_n(s')\pi(s_m,s')\mathrm{d}s' = f(s_m) \text{ for } m = 1, 2, \ldots, N \tag{2.82}$$

The technique for obtaining Eq. (2.82) is called the *point-matching technique*.
    Eq. (2.82) is reduced to

$$\sum_n I_n \int_{\Delta S_n} \pi(s_m,s')\mathrm{d}s' = f(s_m) \text{ for } m = 1, 2, \ldots, N \tag{2.83}$$

where the following expansion functions are chosen

$$J_n(s') = \begin{cases} 1 & \text{for } s' \in \Delta s_n \\ 0 & \text{otherwise} \end{cases} \tag{2.84}$$

Equation (2.83) can be transformed into a matrix form that is similar to Eq. (2.49). Thus, the current distribution $I(s')$ can be determined.

2.6 INTEGRAL EQUATION N1 FOR AN ARBITRARILY SHAPED WIRE ANTENNA: CLOSED KERNEL EXPRESSION

The integral equation (2.72) becomes a powerful analysis tool for an antenna whose shape is expressed by a simple function and, hence, the integrals and derivatives in the kernel, $\pi_1 - \pi_2 - \pi_3$, are readily manipulated. However, if the antenna has a complicated structure, the kernel calculations require cumbersome manipulations. To simplify the programming of the solver and reduce the computing time, it is desirable to derive an integral equation with a closed-form kernel (kernel with a simple algebraic form having no derivatives and no integrals). Such an integral equation has been derived by Nakano [5,11,12]:

$$\sum_j \int_{\Delta j} I_j(s')\pi_{ij}(s_i, s'_j)ds'_j = B\cos\beta(d_i + s_i) - \frac{jV_0}{2Z_0}\sin\beta|d_i + s_i| \tag{2.85}$$

where the kernel is

$$\pi_{ij}(s_i, s'_j) = G_{ij}(s_i, s'_j)\hat{\mathbf{s}}_i \bullet \hat{\mathbf{s}}'_j - \eta_{iij}(s_i, s_i, s'_j) + \eta_{iij}(0, s_i, s'_j)$$
$$- \sum_{k=1}^{i-1}\left[G_{kj}(e_k, s'_j)(\hat{\mathbf{s}}_{k+1} - \hat{\mathbf{s}}_k) \bullet \hat{\mathbf{s}}'_j\cos\beta D_{ik}(s_i, e_k) + \eta_{kij}(e_k, s_i, s'_j) - \eta_{kij}(0, s_i, s'_j)\right] \tag{2.86}$$

This integral equation is derived by subdividing an arbitrarily shaped wire into numerous elements that can be regarded as linear, as shown in Fig. 2.5. The notation in Eq. (2.86) is as follows: d_i is the distance along the antenna arm from the feed gap to the starting point of the ith subdivision; s_i and s'_j are the distances along the antenna arm from the starting points of the ith and jth subdivisions to the observation and source points, respectively $(s_i = s - d_i, \ s'_j = s' - d_j)$; $\hat{\mathbf{s}}_i$ and $\hat{\mathbf{s}}'_j$ are the tangential unit vectors at the points located at distances s_i and s'_j, respectively; Δ_j is the jth subdivision ranging from d_j to $d_j + e_j$; $G_{ij}(s_i, s'_j)$ is the free-space Green's function; e_k is the length of the kth subdivision; and $\eta_{kij}(\xi_k, s_i, s'_j)$ is defined as

$$\eta_{kij}(\xi_k, s_i, s'_j) = [\mathbf{Q}_{kj}(\xi_k, s'_j) \bullet \hat{\mathbf{s}}_k G_{kj}(\xi_k, s'_j)\cos\beta D_{ik}(s_i, \xi_k)$$
$$+ j\frac{1}{4\pi}e^{-j\beta r_{kj}(\xi_k, s'_j)}\sin\beta D_{ik}(s_i, \xi_k)]g_{kj}(\xi_k, s'_j) \tag{2.87}$$

where

$$D_{ik}(s_i, \xi_k) = d_i + s_i - d_k - \xi_k \tag{2.88}$$

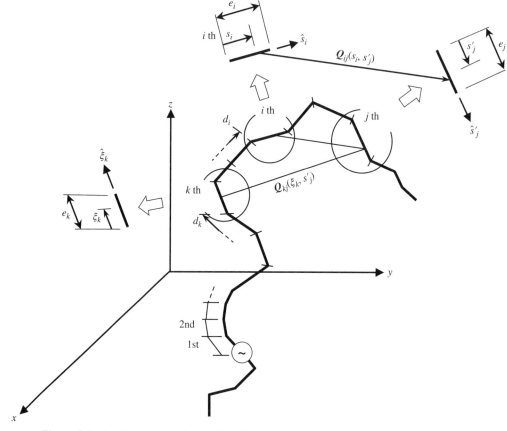

Figure 2.5   Coordinate system for an arbitrarily shaped wire.

and

$$
g_{kj}\left(\xi_k, s'_j\right) = \frac{\mathbf{Q}_{kj}(\xi_k, s'_j) \cdot \hat{\mathbf{s}}'_j - \left[\mathbf{Q}_{kj}(\xi_k, s'_j) \cdot \hat{\mathbf{s}}_k\right]\left(\hat{\mathbf{s}}_k \cdot \hat{\mathbf{s}}'_j\right)}{\left|\mathbf{Q}_{kj}(\xi_k, s'_j)\right|^2 - \left[\mathbf{Q}_{kj}(\xi_k, s'_j) \cdot \hat{\mathbf{s}}_k\right]^2 + a^2}
\tag{2.89}
$$

The $\mathbf{Q}_{kj}(\xi_k, s'_j)$ in Eq. (2.89) is a vector that extends from a point at distance ξ_k on the kth subdivision axis to a point at distance s'_j on the jth subdivision axis. a is the wire radius. Note that $g_{kj}\left(\xi_k, s'_j\right) = g_{kj}\left(0, s'_j\right)$.

From Eq. (2.87), $\eta_{iij}(s_i, s_i, s'_j)$ and $\eta_{iij}(0, s_i, s'_j)$ are given as

$$
\eta_{iij}(s_i, s_i, s'_j) = \left[\mathbf{Q}_{ij}(s_i, s'_j) \cdot \hat{\mathbf{s}}_i\, G_{ij}(s_i, s'_j)\right] g_{ij}(s_i, s'_j)
\tag{2.90}
$$

$$
\eta_{iij}(0, s_i, s'_j) = \left[\mathbf{Q}_{ij}(0, s'_j) \cdot \hat{\mathbf{s}}_i\, G_{ij}(0, s'_j)\cos\beta\, s_i + j\frac{1}{4\pi}\mathrm{e}^{-j\beta\, r_{ij}(0,s'_j)}\sin\beta s_i\right] g_{ij}(0, s'_j) \tag{2.91}
$$

It is emphasized that $\pi_{ij}\left(s_i, s'_j\right)$ leads to simplified programming of the solver for integral equation (2.85), where the main task is to specify the coordinates of the starting and ending points of the subdivisions of the antenna arm.

Exercise

There are two ways to derive Eq. (2.85) characterized by the closed kernel of Eq. (2.86). One way is to use the continuity of vector potentials and the continuity of scalar potentials at each point where sequential subdivisions of the antenna arm are combined [13]. The other is to use Mei's integral equation. Using the latter, obtain closed form expressions for π_2 and π_3.

Answer Using the antenna arm elements, which can be regarded as linear, π_2 of Eq. (2.70) is readily reduced to

$$\pi_2 = \sum_{k=1}^{i-1} G_{kj}(e_k, s'_j)(\hat{s}_{k+1} - \hat{s}_k) \cdot \hat{s}'_j \cos\beta D_{ik}(s_i, e_k) \tag{2.92}$$

where

$$G_{kj}(e_k, s'_j) = \frac{1}{4\pi} \frac{e^{-j\beta r_{kj}(e_k, s'_j)}}{r_{kj}(e_k, s'_j)} \tag{2.93}$$

and

$$D_{ik}(s_i, e_k) = d_i + s_i - d_k - e_k \tag{2.94}$$

The distance $r_{kj}\left(e_k, s'_j\right)$ in the denominator of Eq. (2.93) can be expressed as

$$r_{kj}(e_k, s'_j) = \left[\left|Q_{kj}(e_k, s'_j)\right|^2 + a^2\right]^{\frac{1}{2}} \tag{2.95}$$

where $\mathbf{Q}_{kj}$ is, by definition,

$$\mathbf{Q}_{kj}\left(e_k, s'_j\right) = -e_k\hat{s}_k + \mathbf{Q}_{kj}(0,0) + s'_j\hat{s}'_j \tag{2.96}$$

Next, derive a closed-form expression for π_3. For this, Eq. (2.71) is separated into two groups:

$$\pi_3 = \left(\sum_{k=1}^{i-1} \gamma_k\right) + \gamma_i \tag{2.97}$$

where γ_k and γ_i are defined as

$$\gamma_k = \int_0^{e_k} \left[\frac{\partial G_{kj}\left(\xi_k, s'_j\right)}{\partial \xi_k}\left(\hat{s}_k \cdot \hat{s}'_j\right) + \frac{\partial G_{kj}\left(\xi_k, s'_j\right)}{\partial s'_j}\right] \cos\beta D_{ik}(s_i, \xi_k)\mathrm{d}\xi_k \tag{2.98}$$

$$\gamma_i = \int_0^{s_i} \left[\frac{\partial G_{ij}\left(\xi_i, s'_j\right)}{\partial \xi_i}\left(\hat{s}_i \cdot \hat{s}'_j\right) + \frac{\partial G_{ij}\left(\xi_i, s'_j\right)}{\partial s'_j}\right] \cos\beta D_{ii}(s_i, \xi_i)\mathrm{d}\xi_i \tag{2.99}$$

The γ_k and γ_i have the same form, and hence, first, γ_k is simplified and then the γ_i is expressed using the simplified γ_k.

Using integration by parts, Eq. (2.98) is transformed into

$$\gamma_k = (\hat{\mathbf{s}}_k \cdot \hat{\mathbf{s}}_j')[G_{kj}(e_k, s_j')\cos\beta D_{ik}(s_i, e_k)$$
$$- G_{kj}(0, s_j')\cos\beta D_{ik}(s_i, 0)] + \frac{\partial}{\partial s_j'}\int_0^{e_k} G_{kj}(\xi_k, s_j')\cos\beta D_{ik}(s_i, \xi_k)d\xi_k$$
$$- (\hat{\mathbf{s}}_k \cdot \hat{\mathbf{s}}_j')\beta\int_0^{e_k} G_{kj}(\xi_k, s_j')\sin\beta D_{ik}(s_i, \xi_k)d\xi_k \tag{2.100}$$

The first integral term of the above equation, defined as $K_{\cos}$, is written as

$$K_{\cos} \equiv \int_0^{e_k} G_{kj}(\xi_k, s_j')\cos\beta D_{ik}(s_i, \xi_k)d\xi_k$$
$$= \frac{1}{2}\int_0^{e_k} \frac{e^{-j\beta[r_{kj}(\xi_k, s_j') - D_{ik}(s_i, \xi_k)]}}{4\pi r_{kj}(\xi_k, s_j')}d\xi_k \tag{2.101}$$
$$+ \frac{1}{2}\int_0^{e_k} \frac{e^{-j\beta[r_{kj}(\xi_k, s_j') + D_{ik}(s_i, \xi_k)]}}{4\pi r_{kj}(\xi_k, s_j')}d\xi_k$$

The powers of the exponentials in the numerator of Eq. (2.101) are expressed as

$$r_{kj}(\xi_k, s_j') - D_{ik}(s_i, \xi_k) = m_{kj}^- + [\mathbf{Q}_{kj}(0, s_j') \cdot \hat{\mathbf{s}}_k - D_{ik}(s_i, 0)] \tag{2.102}$$
$$r_{kj}(\xi_k, s_j') + D_{ik}(s_i, \xi_k) = m_{kj}^+ - [\mathbf{Q}_{kj}(0, s_j') \cdot \hat{\mathbf{s}}_k - D_{ik}(s_i, 0)] \tag{2.103}$$

where

$$m_{kj}^+ \equiv r_{kj}(\xi_k, s_j') + \mathbf{Q}_{kj}(\xi_k, s_j') \cdot \hat{\mathbf{s}}_k \tag{2.104}$$
$$m_{kj}^- \equiv r_{kj}(\xi_k, s_j') - \mathbf{Q}_{kj}(\xi_k, s_j') \cdot \hat{\mathbf{s}}_k \tag{2.105}$$

Then, Eq. (2.101) becomes

$$K_{\cos} = \frac{1}{2}e^{-j\beta C_{kj}}\int_0^{e_k} \frac{e^{-j\beta\, m_{kj}^-}}{4\pi r_{kj}(\xi_k, s_j')}d\xi_k$$
$$+ \frac{1}{2}e^{+j\beta C_{kj}}\int_0^{e_k} \frac{e^{-j\beta\, m_{kj}^+}}{4\pi r_{kj}(\xi_k, s_j')}d\xi_k \tag{2.106}$$

where C_{kj} is defined as

$$C_{kj} \equiv \mathbf{Q}_{kj}(0, s_j') \cdot \hat{\mathbf{s}}_k - D_{ik}(s_i, 0) \tag{2.107}$$

After some manipulations, $K_{\cos}$ becomes

$$K_{\cos} = \frac{1}{4\pi}(\Psi^- - \Psi^+) \tag{2.108}$$

where

$$\Psi^+ = \frac{1}{2}e^{+j\beta C_{kj}}\int_{U_{kj}^+(0,s_j')}^{U_{kj}^+(e_k,s_j')} \frac{e^{-j\beta\, m_{kj}^+}}{m_{kj}^+}dm_{kj}^+ \tag{2.109}$$

$$\Psi^- = \frac{1}{2}e^{-j\beta C_{kj}}\int_{U_{kj}^-(0,s_j')}^{U_{kj}^-(e_k,s_j')} \frac{e^{-j\beta\, m_{kj}^-}}{m_{kj}^-}dm_{kj}^- \tag{2.110}$$

where the upper and lower limits in the preceding integrations are given as

$$U_{kj}^{+}(e_k, s_j') = r_{kj}(e_k, s_j') + \mathbf{Q}_{kj}(e_k, s_j') \bullet \hat{\mathbf{s}}_k \tag{2.111}$$

$$U_{kj}^{-}(e_k, s_j') = r_{kj}(e_k, s_j') - \mathbf{Q}_{kj}(e_k, s_j') \bullet \hat{\mathbf{s}}_k \tag{2.112}$$

$$U_{kj}^{+}(0, s_j') = r_{kj}(0, s_j') + \mathbf{Q}_{kj}(0, s_j') \bullet \hat{\mathbf{s}}_k \tag{2.113}$$

$$U_{kj}^{-}(0, s_j') = r_{kj}(0, s_j') - \mathbf{Q}_{kj}(0, s_j') \bullet \hat{\mathbf{s}}_k \tag{2.114}$$

Note that the second integral term in Eq. (2.100), defined as L_{sin}, is similarly calculated to be

$$L_{\sin} \equiv \int_0^{e_k} G_{kj}\left(\xi_k, s_j'\right) \sin \beta \, D_{ik}(s_i, \xi_k) d\xi_k = \frac{1}{j4\pi}(\Psi^- + \Psi^+) \tag{2.115}$$

Substituting Eq. (2.108) and Eq. (2.115) into Eq. (2.100) leads to

$$\gamma_k = (\hat{\mathbf{s}}_k \bullet \hat{\mathbf{s}}_j')[G_{kj}(e_k, s_j')\cos\beta D_{ik}(s_i, e_k) - G_{kj}(0, s_j')\cos\beta D_{ik}(s_i, 0)]$$
$$+ \frac{1}{4\pi} \frac{\partial}{\partial s_j'}(\Psi^- - \Psi^+) + j\frac{\beta}{4\pi}(\hat{\mathbf{s}}_k \bullet \hat{\mathbf{s}}_j')(\Psi^- + \Psi^+) \tag{2.116}$$

After differentiating Ψ^- and Ψ^+ with respect to s_j', γ_k becomes

$$\gamma_k = \eta_{kij}(e_k, s_i, s_j') - \eta_{kij}(0, s_i, s_j') \tag{2.117}$$

where η_{kij} is subject to the definition of Eq. (2.87). Note that the γ_i term of Eq. (2.99) is easily obtained by changing the subscript k of Eq. (2.117) to i and changing e_k to s_i:

$$\gamma_i = \eta_{iij}\left(s_i, s_i, s_j'\right) - \eta_{iij}\left(0, s_i, s_j'\right) \tag{2.118}$$

Substitution of Eq. (2.117) and Eq. (2.118) into Eq. (2.97) yields

$$\pi_3 = \eta_{iij}\left(s_i, s_i, s_j'\right) - \eta_{iij}\left(0, s_i, s_j'\right) + \sum_{k=1}^{i-1}\left[\eta_{kij}\left(e_k, s_i, s_j'\right) - \eta_{kij}\left(0, s_i, s_j'\right)\right] \tag{2.119}$$

It is readily found that substituting Eqs. (2.69), (2.92), and (2.119) into $\pi_1 - \pi_2 - \pi_3$ leads to the kernel $\pi_{ij}\left(s_i, s_j'\right)$ of Eq. (2.86). ∎

2.7 INTEGRAL EQUATIONS N2 AND N3 FOR AN ANTENNA SYSTEM COMPOSED OF AN ARBITRARILY SHAPED WIRE AND AN ARBITRARILY SHAPED APERTURE AND THEIR MoM TRANSFORMATION

Figure 2.6 depicts an antenna system where a plane wave $\left(\mathbf{E}^{in}, \mathbf{H}^{in}\right)$ passes through a narrow aperture in a thin planar conducting plane (perfect electric conductor screen), and excites a thin wire behind the plane. Both the aperture and wire have arbitrary shape [13,14].

The $-z$ region of this antenna system is formulated using Fig. 2.7. Based on Fig. 2.7c, the total magnetic field $\mathbf{H}_{z-}$ in the $-z$ region is given by

$$\mathbf{H}_{z-} = \mathbf{H}_{sc} - j\omega\mathbf{F} + \nabla\frac{\nabla \bullet \mathbf{F}}{j\omega\mu\varepsilon} \tag{2.120}$$

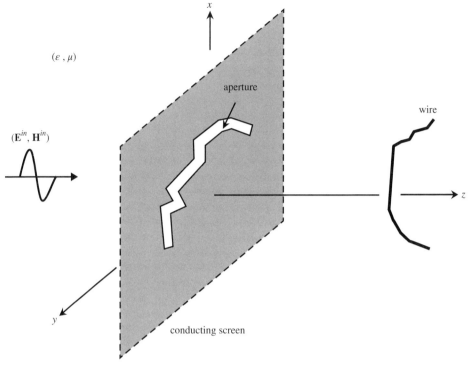

Figure 2.6   Antenna system composed of an arbitrarily shaped wire and an arbitrarily shaped aperture.

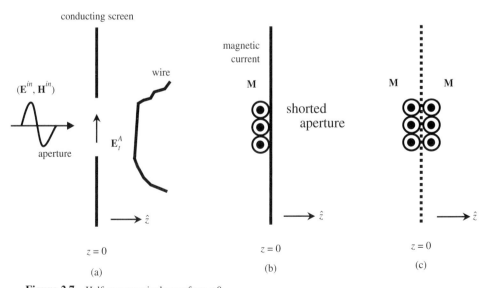

Figure 2.7   Half-space equivalences for $z < 0$.

where **F** is the electric vector potential:

$$\mathbf{F}(\mathbf{r}) = \frac{\varepsilon}{4\pi} \int_{aperture} 2\mathbf{M}(\mathbf{r}')P(\mathbf{r}, \mathbf{r}')ds' \qquad (2.121)$$

with

$$P(\mathbf{r}, \mathbf{r}') = \frac{e^{-j\beta|\mathbf{r}-\mathbf{r}'|}}{|\mathbf{r} - \mathbf{r}'|} \qquad (2.122)$$

The $\mathbf{H}_{sc}$ in Eq. (2.120) is the so-called short-circuit magnetic field, defined as $\mathbf{H}_{sc} = \mathbf{H}^{in} + \mathbf{H}^{r}$, where $\mathbf{H}^{r}$ is the magnetic field reflected from the short-circuit plane (screen). The other terms are defined as $\beta = \omega\sqrt{\mu\varepsilon}$ and $\mathbf{M} = -\mathbf{E}_t^A \times \hat{\mathbf{z}}$, with $\mathbf{E}_t^A$ being the electric field tangential to the aperture.

The $+z$ region of this antenna system is formulated using Fig. 2.8, where Fig. 2.8b illustrates the conducting plane with the short-circuit aperture over which a magnetic current flows. The conducting plane in this figure is removed in Fig. 2.8c by the inclusion of the images of the wire and the magnetic current.

The total magnetic field $\mathbf{H}_{z+}$ in the $+z$ region is written as

$$\mathbf{H}_{z+} = j\omega\mathbf{F} - \nabla\frac{\nabla \bullet \mathbf{F}}{j\omega\mu\varepsilon} + \frac{1}{\mu}\nabla \times \mathbf{A} \qquad (2.123)$$

where **F** is defined in Eq. (2.121), and **A** is the magnetic vector potential generated by wire current **I** across length L_a plus its image:

$$\mathbf{A} = \frac{\mu}{4\pi} \int_{L_a} \mathbf{I}(\mathbf{r}'_W)P(\mathbf{r}, \mathbf{r}'_W)dl' + \frac{\mu}{4\pi} \int_{L_a} \mathbf{I}(\mathbf{r}'_i)P(\mathbf{r}, \mathbf{r}'_i)dl' \qquad (2.124)$$

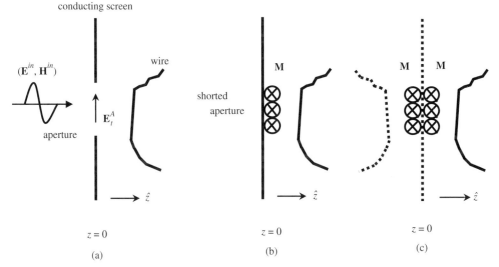

Figure 2.8   Half-space equivalences for $z > 0$.

where $\mathbf{r}'_W$ denotes a vector locating a source point on the wire, and $\mathbf{r}'_i$ denotes a vector locating a source point on the wire image.

The continuity requirement of the tangential magnetic field at the boundary $z = 0$ leads to

$$-\frac{1}{2\mu}(\nabla \times \mathbf{A}) \bullet \hat{\mathbf{s}} + \frac{1}{j\omega\mu\varepsilon}\left(\beta^2 \mathbf{F} + \nabla\nabla \bullet \mathbf{F}\right) \bullet \hat{\mathbf{s}} = -\frac{1}{2}\mathbf{H}_{sc} \bullet \hat{\mathbf{s}} \tag{2.125}$$

where $\hat{\mathbf{s}}$ is a tangential unit vector at an observation point in the aperture. In addition, the boundary condition on the wire surface [i.e., the total electric field in the wire-axis direction specified by a unit vector $\hat{\mathbf{s}}$ is zero] leads to

$$\left(\mathbf{E}_{w,ap} + \mathbf{E}_{w,im} + \mathbf{E}_{w,so}\right) \bullet \hat{\mathbf{s}} = 0 \tag{2.126}$$

where $\mathbf{E}_{w,ap}$ is the field due to the aperture, $\mathbf{E}_{w,im}$ is the field due to the wire image, and $\mathbf{E}_{w,so}$ is the field due to the equivalent sources on the wire. The terms in Eq. (2.126) are

$$\mathbf{E}_{w,ap} = \frac{1}{\varepsilon}\nabla \times \mathbf{F} \tag{2.127}$$

$$\mathbf{E}_{w,so} + \mathbf{E}_{w,im} = -j\omega\mathbf{A} + \nabla\frac{\nabla \bullet \mathbf{A}}{j\omega\mu\varepsilon} \tag{2.128}$$

Eqs. (2.125) and (2.126) form integral equations for wire current $\mathbf{I}$ and aperture magnetic current $\mathbf{M}$. A pair of these equations is labeled as *integral equation N3*, or simply, *equation N3*, for a system composed of an arbitrarily shaped wire and aperture. These Eqs. (2.125) and (2.126) have been reduced to the following Eqs. (2.129) and (2.130), respectively, by Nakano [5], where no derivatives and no integrals are involved.

$$\sum_n I_n h_{n:wire} + \sum_n M_n h_{n:aperture} = -\frac{1}{2}\mathbf{H}_{sc} \bullet \hat{\mathbf{s}} \tag{2.129}$$

$$\sum_n I_n e_{n:wire} + \sum_n M_n e_{n:aperture} = 0 \tag{2.130}$$

where

$$I_n h_{n:wire} = -I_n\frac{j}{8\pi}\left(\frac{\Phi_{n-1}}{\rho_{n-1}\sin\beta e_{n-1}} + \frac{\Phi_n}{\rho_n\sin\beta e_n}\right) \bullet \hat{\mathbf{s}} \tag{2.131}$$

$$M_n h_{n:aperture} = -M_n\frac{j}{2\pi Z_0}\left[\left(\frac{P_{n-1}}{\rho_{n-1}\sin\beta e_{n-1}} + \frac{P_n}{\rho_n\sin\beta e_n}\right) - \left(\frac{Z_{n-1}}{\sin\beta e_{n-1}} + \frac{Z_n}{\sin\beta e_n}\right)\right] \bullet \hat{\mathbf{s}} \tag{2.132}$$

$$I_n e_{n:wire} = -I_n j30\left[\left(\frac{P_{n-1}}{\rho_{n-1}\sin\beta e_{n-1}} + \frac{P_n}{\rho_n\sin\beta e_n}\right) - \left(\frac{Z_{n-1}}{\sin\beta e_{n-1}} + \frac{Z_n}{\sin\beta e_n}\right)\right] \bullet \hat{\mathbf{s}} \tag{2.133}$$

$$M_n e_{n:aperture} = M_n\frac{j}{2\pi}\left(\frac{\Phi_{n-1}}{\rho_{n-1}\sin\beta e_{n-1}} + \frac{\Phi_n}{\rho_n\sin\beta e_n}\right) \bullet \hat{\mathbf{s}} \tag{2.134}$$

with vectors

$$\mathbf{\Phi}_{n-1} = \left[\left(j\cos\theta_U^{n-1}\sin\beta e_{n-1} - \cos\beta e_{n-1}\right)\mathrm{e}^{-j\beta\,R_U^{n-1}} + \mathrm{e}^{-j\beta\,R_L^{n-1}}\right]\hat{\mathbf{\phi}}_{n-1} \tag{2.135}$$

$$\mathbf{\Phi}_n = \left[-\left(j\cos\theta_L^n\sin\beta e_n + \cos\beta e_n\right)\mathrm{e}^{-j\beta\,R_L^n} + \mathrm{e}^{-j\beta\,R_U^n}\right]\hat{\mathbf{\phi}}_n \tag{2.136}$$

$$\mathbf{P}_{n-1} = \left(\cos\beta e_{n-1}\cos\theta_U^{n-1}\cdot\mathrm{e}^{-j\beta\,R_U^{n-1}} - \cos\theta_L^{n-1}\mathrm{e}^{-j\beta\,R_L^{n-1}} - j\sin\beta e_{n-1}\mathrm{e}^{-j\beta\,R_U^{n-1}}\right)\hat{\mathbf{\rho}}_{n-1} \tag{2.137}$$

$$\mathbf{P}_n = \left(\cos\beta e_n\cos\theta_L^n\mathrm{e}^{-j\beta\,R_L^n} - \cos\theta_U^n\mathrm{e}^{-j\beta\,R_U^n} + j\sin\beta e_n\mathrm{e}^{-j\beta\,R_L^n}\right)\hat{\mathbf{\rho}}_n \tag{2.138}$$

$$\mathbf{Z}_{n-1} = \left(\cos\beta e_{n-1}\frac{\mathrm{e}^{-j\beta\,R_U^{n-1}}}{R_U^{n-1}} - \frac{\mathrm{e}^{-j\beta\,R_L^{n-1}}}{R_L^{n-1}}\right)\hat{\mathbf{z}}_{n-1} \tag{2.139}$$

$$\mathbf{Z}_n = \left(-\frac{\mathrm{e}^{-j\beta\,R_U^n}}{R_U^n} + \cos\beta e_n\frac{\mathrm{e}^{-j\beta\,R_L^n}}{R_L^n}\right)\hat{\mathbf{z}}_n \tag{2.140}$$

Equation (2.131) is formulated using the following three steps. First, the arbitrarily shaped wire is subdivided into numerous segments, each regarded as being linear. Figure 2.9 depicts the consecutive subdivisions $n-1$ (of length e_{n-1}) and n (of length e_n), where the local cylindrical coordinates $(\rho_{n-1},\phi_{n-1},z_{n-1})$ and (ρ_n,ϕ_n,z_n) are adopted for subdivisions $n-1$ and n with unit vectors $(\hat{\mathbf{\rho}}_{n-1},\hat{\mathbf{\phi}}_{n-1},\hat{\mathbf{z}}_{n-1})$ and $(\hat{\mathbf{\rho}}_n,\hat{\mathbf{\phi}}_n,\hat{\mathbf{z}}_n)$, respectively. The distances (R_L^{n-1},R_U^{n-1}) and (R_L^n,R_U^n) are defined in this figure, together with angles $(\theta_L^{n-1},\theta_U^{n-1})$ and (θ_L^n,θ_U^n).

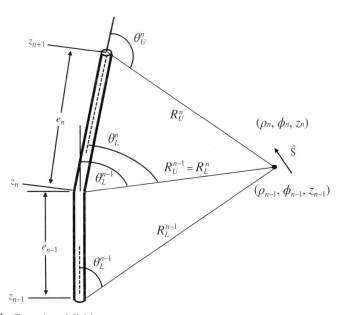

Figure 2.9 Two wire subdivisions.

Second, the current on subdivisions $n-1$ and n is expanded using piecewise sinusoidal functions

$$I(z') = \begin{cases} I_n \dfrac{\sin \beta(z' - z_{n-1})}{\sin \beta\, e_{n-1}} & for\; z_{n-1} \leq z' \leq z_n \\ I_n \dfrac{\sin \beta(z_{n+1} - z')}{\sin \beta\, e_n} & for\; z_n \leq z' \leq z_{n+1} \end{cases} \tag{2.141}$$

Third, using local cylindrical coordinates and the expansion functions in Eq. (2.141), the first term on the left side of Eq. (2.125) is calculated. Thus, $I_n\, h_{n:\text{wire}}$ [Eq. (2.131), i.e., the tangential component of a magnetic field resulting from these two *wire* subdivisions, $n-1$ and n] is obtained.

Similarly, $M_n\, h_{n:\,\text{aperture}}$ [Eq. (2.132)] is obtained from the second term on the left side of Eq. (2.125). For this, the narrow aperture is subdivided into numerous segments (each regarded as being linear) and the magnetic current **M** is expanded using the piecewise sinusoidal functions in Eq. (2.141), where I_n is replaced by M_n.

$I_n\, e_{n:\text{wire}}$ [Eq. (2.133)] and $M_n\, e_{n:\text{aperture}}$ [Eq. (2.134)] are also obtained from Eq. (2.128) and Eq. (2.127), respectively, using local cylindrical coordinates for the subdivisions of the wire and aperture. It is emphasized that the obtained results of $I_n\, h_{n:\text{wire}}$, $M_n\, h_{n:\,\text{aperture}}$, $I_n\, e_{n:\text{wire}}$, and $M_n\, e_{n:\text{aperture}}$ are all simplified, in a form without derivatives and integrals. The expansion function coefficients, I_n and M_n, are determined by standard matrix methods after weighting Eqs. (2.129) and (2.130) with properly chosen functions (e.g., piecewise sinusoidal functions).

Note that, when a system does not have wire conductors, Eq. (2.129) is reduced to

$$\sum_n M_n\, \dot{h}_{n:\text{aperture}} = -\frac{1}{2}\mathbf{H}_{sc} \bullet \hat{\mathbf{s}} \tag{2.142}$$

This equation is used to analyze a narrow aperture antenna. Conversely, when a system does not have apertures, Eq. (2.130) is reduced to

$$\sum_n I_n e_{n:\text{wire}} = 0 \tag{2.143}$$

This equation is used to analyze a wire antenna above a conducting plane. A pair of integral equations yielding Eqs. (2.142) and (2.143) is labeled as *integral equation N2*, which is a special case of integral equation N3.

Exercise

Figure 2.10 shows a straight-wire subdivision. Express the first term on the left side in Eq. (2.125) in a form without derivatives and integrals, under the condition that the current $I(z')$ on the wire is distributed as follows:

$$I(z') = \frac{I_L \sin\beta(z' - z_L) + I_U \sin\beta(z_U - z')}{\sin\beta(z_U - z_L)} \tag{2.144}$$

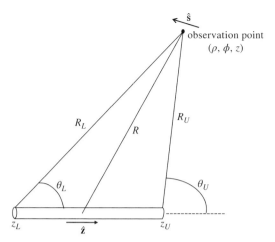

Figure 2.10   A straight-wire subdivision.

Answer   Using the local cylindrical coordinates (ρ, ϕ, z) with unit vectors $(\hat{\rho}, \hat{\phi}, \hat{z})$, and letting the z-axis coincide with the subdivision axis, the first term is written as

$$-\frac{1}{2\mu}(\nabla \times \mathbf{A}) \bullet \hat{s} = \frac{1}{2\mu}\frac{\partial A_z}{\partial \rho}\hat{\phi} \bullet \hat{s} \qquad (2.145)$$

where A_z is the z-axis component of magnetic vector potential $\mathbf{A}$. Equation (2.145) is separated into two terms:

$$\frac{1}{2\mu} \cdot \frac{\partial A_z}{\partial \rho}\hat{\phi} \bullet \hat{s} = f_\phi(I_L)\hat{\phi} \bullet \hat{s} + g_\phi(I_U)\hat{\phi} \bullet \hat{s} \qquad (2.146)$$

where

$$f_\phi(I_L) = -\frac{1}{j16\pi} \cdot \frac{I_L}{\sin \beta e} \cdot \frac{\partial}{\partial \rho} \int_{Z_L}^{Z_U} \left(\frac{e^{-j\beta[R-(z_L-z')]}}{R} - \frac{e^{-j\beta[R+(z_L-z')]}}{R} \right) dz' \qquad (2.147)$$

$$g_\phi(I_U) = \frac{1}{j16\pi} \cdot \frac{I_U}{\sin \beta e} \cdot \frac{\partial}{\partial \rho} \int_{Z_L}^{Z_U} \left(\frac{e^{-j\beta[R-(z_U-z')]}}{R} - \frac{e^{-j\beta[R+(z_U-z')]}}{R} \right) dz' \qquad (2.148)$$

with

$$R = \left[(z - z')^2 + \rho^2 \right]^{1/2} \qquad (2.149)$$

and $e = z_U - z_L$. When z_L and z_U are designated as α, the derivatives with respect to ρ in Eqs. (2.147) and (2.148) are given as

$$\frac{\partial}{\partial \rho} \int_{Z_L}^{Z_U} \frac{e^{-j\beta[R-(\alpha-z')]}}{R} dz' = \frac{e^{-j\beta[R_U+z_U-\alpha]}}{U^+}\frac{\rho}{R_U} - \frac{e^{-j\beta[R_L+z_L-\alpha]}}{L^+}\frac{\rho}{R_L} \qquad (2.150)$$

$$\frac{\partial}{\partial \rho} \int_{Z_L}^{Z_U} \frac{e^{-j\beta[R+(\alpha-z')]}}{R} dz' = -\frac{e^{-j\beta[R_U+\alpha-z_U]}}{U^-}\frac{\rho}{R_U} + \frac{e^{-j\beta[R_L+\alpha-z_L]}}{L^-}\frac{\rho}{R_L} \qquad (2.151)$$

where

$$U^+ = R_U + z_U - z \qquad (2.152a)$$

$$L^+ = R_L + z_L - z \qquad (2.152b)$$

$$U^- = R_U - z_U + z \qquad (2.153a)$$

$$L^- = R_L - z_L + z \qquad (2.153b)$$

The R_L and R_U in Eqs. (2.150) and (2.151) are the distance from the starting and ending points of the straight-wire subdivision to the observation point, respectively, as shown in Fig. 2.10.

Using Eqs. (2.150) and (2.151), Eqs. (2.147) and (2.148) are transformed into

$$f_\phi(I_L) = -I_L\frac{j}{8\pi\rho} \cdot \frac{1}{\sin\beta e}\left[(j\cos\theta_U \cdot \sin\beta e - \cos\beta e)e^{-j\beta R_U} + e^{-j\beta R_L}\right] \qquad (2.154)$$

and

$$g_\phi(I_U) = -I_U\frac{j}{8\pi\rho} \cdot \frac{1}{\sin\beta e}\left[-(j\cos\theta_L \cdot \sin\beta e + \cos\beta e)e^{-j\beta R_L} + e^{-j\beta R_U}\right] \qquad (2.155)$$

where θ_L and θ_U are the angles measured from the z-axis, as shown in Fig. 2.10. Thus, the first term on the left side in Eq. (2.125) is expressed in a form without derivatives and integrals. ∎

2.8 INTEGRAL EQUATION N4 FOR AN ARBITRARILY SHAPED WIRE ANTENNA ON A DIELECTRIC SUBSTRATE BACKED BY A CONDUCTING PLANE AND ITS MoM TRANSFORMATION

Figure 2.11 shows the antenna configuration to be considered here, in which an arbitrarily shaped thin wire is mounted on a dielectric substrate of infinite extent. This dielectric substrate has relative permittivity ε_r and thickness B and is backed by a conductor plane. The arbitrarily shaped wire is assumed to be perfectly

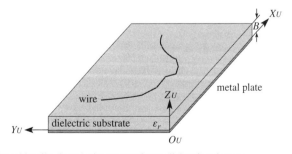

Figure 2.11   An arbitrarily shaped wire mounted on a dielectric substrate.

conducting, and the wire radius a is small, compared with the operating wavelength, so that only the wire-axis component of the current contributes to the antenna characteristics. The purpose of this section is to formulate the current along the arbitrarily shaped wire, applying the method of moments (MoM).

As in the previous section, the arbitrarily shaped wire is subdivided into numerous segments, each regarded as linear. The electric field on the surface of the wire subdivisions, $\mathbf{E}$, is formulated using two steps. The first step is to derive the electric field, $\mathbf{e}$, which is generated by the current on a single subdivision. The second step is to sum all the electric fields generated by the currents on the wire subdivisions: $\mathbf{E} = \sum \mathbf{e}$.

2.8.1 Step I

The electric field $\mathbf{e}$ radiated from current $I(x')\hat{\mathbf{x}}$ flowing on a subdivision of length d shown in Fig. 2.12 is expressed as

$$\mathbf{e} = \int_0^d \left[(k^2 \bar{\mathbf{I}} + \nabla\nabla) \cdot \bar{\bar{\Pi}} \right] \cdot I(x')\hat{\mathbf{x}} dx' \qquad (2.156)$$

where k denotes the phase constant in free space, $\bar{\mathbf{I}} = \hat{\mathbf{x}}\hat{\mathbf{x}} + \hat{\mathbf{y}}\hat{\mathbf{y}} + \hat{\mathbf{z}}\hat{\mathbf{z}}$ is the unit dyadic, and $\bar{\bar{\Pi}}$ is the dyadic Hertz vector potential function. Note that the time dependence, $e^{j\omega t}$, is omitted.

The electric field $\mathbf{e}$ is designated as the unit field. The total field at an observation point on the arbitrarily shaped wire is obtained by summing the unit fields over all the subdivisions, that is, $\mathbf{E} = \sum \mathbf{e}$. The boundary condition at the wire surface, $\mathbf{E} \cdot \hat{\mathbf{t}} = -\mathbf{E}^{in} \cdot \hat{\mathbf{t}}$ (i.e., $E_{tan} = -E_{tan}^{in}$), forms an integral equation for $I(x')$ based on Eq. (2.156). This integral equation is labeled as *integral equation N4*, where $\mathbf{E}^{in}$ is the electric field incident or impressed on the wire and $\hat{\mathbf{t}}$ is the unit vector tangential to the wire.

For simplifying the kernel of integral equation N4, let the current on the subdivision of length d be

$$I(x') = I_L \frac{\sin kx'}{\sin kd} + I_U \frac{\sin k(d - x')}{\sin kd} \qquad (2.157)$$

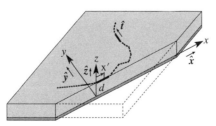

Figure 2.12 The coordinate x' on a subdivision of length d and a unit vector $\hat{\mathbf{t}}$ tangential to the subdivision.

Then, the tangential component of the electric field at an observation point $(x, y, z = B)$ located on any subdivision (with tangential unit vector $\hat{t}$), is written as

$$\mathbf{e} \cdot \hat{t} \equiv e_t(x, y, B) = I_L \frac{1}{\sin kd}(\mathbf{L} \cdot \hat{t}) + I_U \frac{1}{\sin kd}(\mathbf{U} \cdot \hat{t}) \qquad (2.158)$$

where

$$\mathbf{L} = \int_0^d (\Gamma^x \hat{x} + \Gamma^y \hat{y}) \sin kx' \, dx' \qquad (2.159)$$

$$\mathbf{U} = \int_0^d (\Gamma^x \hat{x} + \Gamma^y \hat{y}) \sin k(d - x') \, dx' \qquad (2.160)$$

The definitions for Γ^x and Γ^y in the above equations are

$$\Gamma^x = \frac{\partial^2 \Pi^x}{\partial x^2} + \frac{\partial^2 \Pi^z}{\partial x \partial z} + k^2 \Pi^x \qquad (2.161)$$

$$\Gamma^y = \frac{\partial^2 \Pi^x}{\partial y \partial x} + \frac{\partial^2 \Pi^z}{\partial y \partial z} \qquad (2.162)$$

where Π^x and Π^z are Sommerfeld-type integrals [5,15,16]:

$$\Pi^x = 2u \lim_{z \to B} \int_0^\infty J_0(\rho\lambda) e^{-\mu(z-B)} \frac{\lambda}{D_e(\lambda)} \, d\lambda \qquad (2.163)$$

$$\Pi^z = 2u(1 - \varepsilon_r) \lim_{z \to B} \int_0^\infty \cos\phi J_1(\rho\lambda) e^{-\mu(z-B)} \frac{\lambda^2}{D_e(\lambda) D_m(\lambda)} \, d\lambda \qquad (2.164)$$

In the above equations, $\cos\phi = (x - x')/\rho$, $u = -j/(4\pi\varepsilon_0\omega)$, and $J_0(\rho\lambda)$ and $J_1(\rho\lambda)$ denote the Bessel functions of the first kind of order 0 and 1, respectively, where ρ is the distance between the source point (x', y', B) and the observation point (x, y, B), and

$$D_e(\lambda) = \mu + \mu_e \coth(\mu_e B) \qquad (2.165)$$

$$D_m(\lambda) = \mu\varepsilon_r + \mu_e \tanh(\mu_e B) \qquad (2.166)$$

with

$$\mu = \sqrt{\lambda^2 - k^2} \qquad (2.167)$$

$$\mu_e = \sqrt{\lambda^2 - \varepsilon_r k^2} \qquad (2.168)$$

Using Eqs. (2.159), (2.161), and (2.162), the inner product $\mathbf{L} \cdot \hat{t}$ in Eq. (2.158) is expressed as

$$
\begin{aligned}
\mathbf{L} \cdot \hat{t} &= \int_0^d \left[\frac{\partial}{\partial x} \frac{\partial \Pi^x}{\partial x} (\hat{x} \cdot \hat{t}) + \frac{\partial}{\partial y} \frac{\partial \Pi^x}{\partial x} (\hat{y} \cdot \hat{t}) \right] \sin kx' \, dx' \\
&\quad + \int_0^d \left[\frac{\partial}{\partial x} \frac{\partial \Pi}{\partial x'} (\hat{x} \cdot \hat{t}) + \frac{\partial}{\partial y} \frac{\partial \Pi}{\partial x'} (\hat{y} \cdot \hat{t}) \right] \sin kx' \, dx' + (\hat{x} \cdot \hat{t}) \int_0^d k^2 \Pi^x \sin kx' \, dx' \\
&= \int_0^d \frac{\partial}{\partial t} \left(\frac{\partial \Pi^x}{\partial x} + \frac{\partial \Pi}{\partial x'} \right) \sin kx' \, dx' + (\hat{x} \cdot \hat{t}) \int_0^d k^2 \Pi^x \sin kx' \, dx'
\end{aligned}
$$

$$(2.169)$$

where Π is defined as

$$\Pi = 2u(\varepsilon_r - 1) \lim_{z \to B} \int_0^{\infty} J_0(\rho\lambda) e^{-\mu(z-B)} \frac{\lambda\mu}{D_e(\lambda)D_m(\lambda)} \, d\lambda \qquad (2.170)$$

Similarly, the inner product $\mathbf{U} \cdot \hat{\mathbf{t}}$ in Eq. (2.158) is simplified to

$$\mathbf{U} \cdot \hat{\mathbf{t}} = \int_0^d \frac{\partial}{\partial t} \left(\frac{\partial \Pi^x}{\partial x} + \frac{\partial \Pi}{\partial x'} \right) \sin k(d - x') dx' + (\hat{\mathbf{x}} \cdot \hat{\mathbf{t}}) \int_0^d k^2 \Pi^x \sin k(d - x') dx' \qquad (2.171)$$

2.8.2 Step II

Figure 2.13a shows the wire subdivision in the universal coordinate system (X_U, Y_U, Z_U), where all subdivided segments have length d. Figure 2.13b illustrates the details of the local coordinate system (x_n, y_n, z_n) with unit vectors $(\hat{\mathbf{x}}_n, \hat{\mathbf{y}}_n, \hat{\mathbf{z}}_n)$ for subdivision n. The coordinate x_n coincides with the axis of subdivision n.

Let the current for two consecutive subdivisions $n - 1$ and n be

$$I(x'_{n-1}) = I_n \frac{\sin kx'_{n-1}}{\sin kd} \text{ on subdivision } n - 1 \qquad (2.172a)$$

$$I(x'_n) = I_n \frac{\sin k(d - x'_n)}{\sin kd} \text{ on subdivision } n \qquad (2.172b)$$

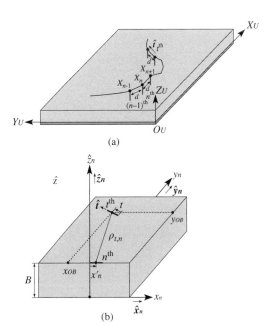

Figure 2.13 Wire subdivisions. (a) Subdivisions $n-1$ and n in the universal coordinate system. (b) Local rectangular coordinate system for subdivision n.

where I_n is the MoM expansion-function coefficient to be determined. Then, using Eq. (2.158), the sum of the tangential components of the electric fields radiated from these consecutive subdivisions $n - 1$ and n is written as

$$e_{n,t} = I_n (\mathbf{e}_n \cdot \hat{\mathbf{t}}) \qquad (2.173)$$

where

$$\mathbf{e}_n \cdot \hat{\mathbf{t}} = \frac{1}{\sin kd} (\mathbf{L}_{n-1} \cdot \hat{\mathbf{t}} + \mathbf{U}_n \cdot \hat{\mathbf{t}}) \qquad (2.174)$$

Adding subscripts $n - 1$ and n to Eq. (2.169) and Eq. (2.171), respectively, $\mathbf{L}_{n-1} \cdot \hat{\mathbf{t}}$ and $\mathbf{U}_n \cdot \hat{\mathbf{t}}$ are

$$\mathbf{L}_{n-1} \cdot \hat{\mathbf{t}} = \int_0^d \frac{\partial}{\partial t} \left(\frac{\partial G_{t,n-1}}{\partial x'_{n-1}} \right) \sin kx'_{n-1} \, dx'_{n-1} + (\hat{\mathbf{x}}_{n-1} \cdot \hat{\mathbf{t}}) \int_0^d k^2 \Pi^x_{t,n-1} \sin kx'_{n-1} \, dx'_{n-1}$$
$$(2.175)$$

$$\mathbf{U}_n \cdot \hat{\mathbf{t}} = \int_0^d \frac{\partial}{\partial t} \left(\frac{\partial G_{t,n}}{\partial x'_n} \right) \sin k(d - x'_n) dx'_n + (\hat{\mathbf{x}}_n \cdot \hat{\mathbf{t}}) \int_0^d k^2 \Pi^x_{t,n} \sin k(d - x'_n) dx'_n \qquad (2.176)$$

where $G_{t,n-1}$ and $G_{t,n}$ are defined as

$$G_{t,j} = -\Pi^x_{t,j} + \Pi_{t,j} \quad (j = n - 1, n) \qquad (2.177)$$

Note that $\Pi^x_{t,j}$ and $\Pi_{t,j}$ in Eq. (2.177) correspond to Π^x in Eq. (2.163) and Π in Eq. (2.170), respectively, and the subscripts t and j indicate that the observation point and source point are located on the subdivisions labeled t and $j(= n - 1, n)$, respectively. Using these subscripts, the distance ρ is written as $\rho_{t,j} = [(x_{OB} - x'_j)^2 + y_{OB}^2]^{1/2}$, where the coordinates of the observation point, x_{OB} and y_{OB}, are those measured using the jth $(j = n - 1, n)$ local coordinate system. Also note that the relation $\partial \Pi^x_{t,j}/\partial x_j = -\partial \Pi^x_{t,j}/\partial x'_j$ is used in Eqs. (2.175) and (2.176).

Finally, the tangential component of the total electric field is obtained by summing Eq. (2.174) over all the currents on the wire subdivisions:

$$E_{\tan} = \frac{1}{\sin kd} \sum_n I_n \left[\int_0^d N_{t,n-1} \sin kx'_{n-1} dx'_{n-1} + \int_0^d N_{t,n} \sin k(d - x'_n) dx'_n \right] \qquad (2.178)$$

where

$$N_{t,j} = \frac{\partial}{\partial t} \left(\frac{\partial G_{t,j}}{\partial x'_j} \right) + k^2 (\hat{x}_j \cdot \hat{t}) \Pi^x_{t,j} \qquad (j = n - 1, n) \qquad (2.179)$$

Equation (2.179) was derived by Nakano [5,16–18]. The coefficients I_n ($n = 1, 2, \ldots$) are solved with impedance matrix elements shown in the following exercise. Note that the MoM formulation in this section can be applied to an arbitrarily shaped narrow *strip* antenna, where wire radius a is replaced by a strip width of $w = 4a$.

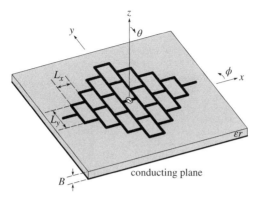

Figure 2.14   Grid array antenna.

Exercise

Obtain the MoM impedance matrix elements based on the boundary condition $-E_{\tan}^{in} = E_{\tan}$, where $E_{\tan}^{in}$ is the tangential component of the electric field incident or impressed on the surface of the arbitrarily shaped wire shown in Fig. 2.11, and $E_{\tan}$ is the electric field given by Eq. (2.178). In addition, obtain the current distribution applying N4 to the grid array shown in Fig. 2.14. Use the parameters shown in Table 2.1, where λ_{12} is the free-space wavelength at frequency 12 GHz.

Answer   Piecewise sinusoidal functions that are the same as the expansion functions are chosen as the weighting functions for the MoM. Applying the weighting functions to both sides of $-E_{\tan}^{in} = E_{\tan}$ leads to an MoM matrix expression of $[V] = [Z][I]$. The impedance matrix element $Z_{m,n}$ is given by

$$Z_{m,n} = \int_0^d \frac{\sin k x_{m-1}}{\sin kd} (\mathbf{e}_n \cdot \hat{\mathbf{x}}_{m-1}) \, dx_{m-1} + \int_0^d \frac{\sin k(d - x_m)}{\sin kd} (\mathbf{e}_n \cdot \hat{\mathbf{x}}_m) \, dx_m \qquad (2.180)$$

where x_{m-1} and x_m are the x-coordinates (distances) in the local coordinate system for consecutive subdivisions $m - 1$ and m, respectively. The inner product terms $(\mathbf{e}_n \cdot \hat{\mathbf{x}}_{m-1})$ and $(\mathbf{e}_n \cdot \hat{\mathbf{x}}_m)$ are $(\mathbf{e}_n \cdot \hat{\mathbf{t}})$ from Eq. (2.174) with substitutions $\hat{\mathbf{t}} = \hat{\mathbf{x}}_{m-1}$ and $\hat{\mathbf{t}} = \hat{\mathbf{x}}_m$, respectively.

Table 2.1   Configuration Parameters for a Grid Array Antenna

Symbol	Value
L_x	$56 \, \Delta_{grid}$
L_y	$112 \, \Delta_{grid}$
B	$8 \, \Delta_{grid}$
ε_r	2.6
ρ	$0.3 \, \Delta_{grid}$
Δ_{grid}	$0.0066 \lambda_{12}$

After some manipulation, Eq. (2.180) is simplified to

$$Z_{m,n} = \left(\frac{k}{\sin kd}\right)^2 (g_{m-1,n-1} + g_{m-1,n} + g_{m,n-1} + g_{m,n}) \tag{2.181}$$

where $g_{i,j}(i = m - 1, m; j = n - 1, n)$ are

$$g_{m-1,j} = \int_0^d (\cos kx_{m-1})[h^{(1)}_{m-1,j}]dx_{m-1}$$

$$+ (\hat{\mathbf{x}}_{m-1} \bullet \hat{\mathbf{x}}_j) \int_0^d (\sin kx_{m-1})[h^{(2)}_{m-1,j}]dx_{m-1}$$

$$+ \int_0^d (\cos kx_{m-1})[h^{(3)}_{m-1,j}]dx_{m-1} \tag{2.182}$$

$$g_{m,j} = -\int_0^d [\cos k(d - x_m)][h^{(1)}_{m,j}]dx_m$$

$$+ (\hat{\mathbf{x}}_m \bullet \hat{\mathbf{x}}_j) \int_0^d [\sin k(d - x_m)][h^{(2)}_{m,j}]dx_m$$

$$- \int_0^d [\cos k(d - x_m)][h^{(3)}_{m,j}]dx_m \tag{2.183}$$

The $h^{(s)}_{i,j}$ ($i = m - 1, m; j = n - 1, n; s = 1, 2, 3$) in Eqs. (2.182) and (2.183) are defined as

$$h^{(1)}_{i,n-1} = -\int_0^d \Pi^x_{i,n-1} \cos kx'_{n-1} dx'_{n-1} \tag{2.184}$$

$$h^{(1)}_{i,n} = \int_0^d \Pi^x_{i,n} \cos k(d - x'_n) dx'_n \tag{2.185}$$

$$h^{(2)}_{i,n-1} = \int_0^d \Pi^x_{i,n-1} \sin kx'_{n-1} dx'_{n-1} \tag{2.186}$$

$$h^{(2)}_{i,n} = \int_0^d \Pi^x_{i,n} \sin k(d - x'_n) dx'_n \tag{2.187}$$

$$h^{(3)}_{i,n-1} = \int_0^d \Pi_{i,n-1} \cos kx'_{n-1} dx'_{n-1} \tag{2.188}$$

$$h^{(3)}_{i,n} = -\int_0^d \Pi_{i,n} \cos k(d - x'_n) dx'_n \tag{2.189}$$

It is found that (i) the second term in Eqs. (2.182) and (2.183) on the right side is dependent on the antenna configuration, that is, the direction of each subdivision; (ii) the third integral in Eqs. (2.182) and (2.183) vanishes when the relative permittivity ε_r is one; (iii) an antenna of arbitrary shape in free space backed by a plane conductor can be analyzed using just the first and second terms of Eqs. (2.182) and (2.183); and (iv) the impedance matrix element has diagonal symmetry, $Z_{m,n} = Z_{n,m}$. The use of this symmetry property reduces the computational burden and reduces the computation time for the [Z] calculation. Further reduction in the computation time for the [Z] calculation is found in Refs 19 and 20.

The current distribution for the grid at 12 GHz, determined by using N4, is shown in Fig. 2.15. The current at this frequency forms standing waves along the grid wires [19]. ■

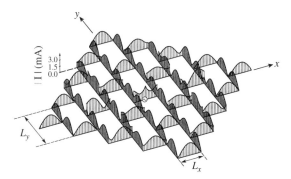

Figure 2.15   Current distribution for the grid determined by using N4.

2.9 INTEGRAL EQUATION N5 FOR AN ARBITRARILY SHAPED WIRE ANTENNA ON A DIELECTRIC HALF-SPACE AND ITS TRANSFORMATION USING A FINITE-DIFFERENCE TECHNIQUE

Figure 2.16 shows a thin wire of arbitrary shape on a dielectric material that occupies the negative-z half-space [21,22]. The dielectric has relative permittivity ε_r and is assumed to be lossless. It is also assumed that the wire is perfectly conducting and the wire radius a is much smaller than the guided wavelength λ_g in the dielectric.

The tangential electric field along the wire antenna, $E_s(s)$, is expressed as

$$
\begin{aligned}
E_s(s) &= \int_0^L \left(\frac{\partial}{\partial s} \frac{\partial}{\partial s'} \left[-\Pi^{(a)s}(s, s') + \Pi^{(a)}(s, s') \right] + k_0^2 \Pi^{(a)s}(s, s') \hat{s}' \bullet \hat{s} \right) I(s') ds' \\
&= -E_{in}(s)
\end{aligned}
\tag{2.190}
$$

where curvilinear coordinates are used for the observation and source points, that is, s and s' are the distance along the antenna arm from the starting point to the observation and source points, respectively, and $\hat{s}$ and $\hat{s}'$ are the tangential unit vectors at

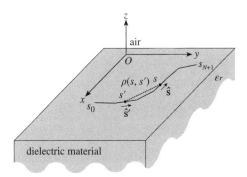

Figure 2.16   Antenna on a dielectric half-space.

these points, respectively. Note that $k_0^2 = \omega^2 \mu_0 \varepsilon_0$. $I(s')$ is the unknown current at s', and L is the antenna length. $E_{in}(s)$ is the tangential component of the electric field incident or impressed on the wire. $\Pi^{(a)s}(s, s')$ and $\Pi^{(a)}(s, s')$ are

$$\Pi^{(a)s}(s, s') = 2u^{(a)} \int_0^\infty J_0[\lambda \rho(s, s')] \times \frac{\lambda}{\sqrt{\lambda^2 - k_0^2} + \sqrt{\lambda^2 - k^2}} \, d\lambda \tag{2.191}$$

$$\Pi^{(a)}(s, s') = 2u^{(a)} \int_0^\infty J_0[\lambda \rho(s, s')] \times \frac{\lambda \sqrt{\lambda^2 - k_0^2} \left(\sqrt{\lambda^2 - k_0^2} - \sqrt{\lambda^2 - k^2} \right)}{k^2 \sqrt{\lambda^2 - k_0^2} + k_0^2 \sqrt{\lambda^2 - k^2}} \, d\lambda \tag{2.192}$$

where J_0 is the Bessel function of the first kind of order zero, and $\rho(s, s')$ is the distance between the observation and source points (see Fig. 2.16). Also, $k^2 = \omega^2 \mu_0 \varepsilon_0 \varepsilon_r$ and $u^{(a)} = -j\omega \mu_0 / 4\pi k_0^2$.

Integral equation (2.190) is called *integral equation N5* [21,22]. Note that integral equation N5 involves double-integral calculations with respect to s' and λ. In addition, N5 has partial derivatives with respect to s and s'. In the following, N5 is transformed to an expression that involves only single-integral calculations without the derivatives, using a wire subdivision technique demonstrated in the previous sections.

The wire is subdivided into $N + 1$ linear segments, each having a length of $d_j = s_{j+1} - s_j (j = 0, 1, 2, \ldots, N)$, where s_{j+1} and s_j are the curvilinear coordinate values (distances) for the ending and starting points of subdivision j, respectively (see Fig. 2.17). Using the condition that the currents at the arm ends are zero and applying a finite difference $\Delta s = \Delta s^+ + \Delta s^-$ for a partial derivative with respect to s, Eq. (2.190) is written as

$$-E_{in}(s) = E_{s1}(s) + E_{s2}(s) \tag{2.193}$$

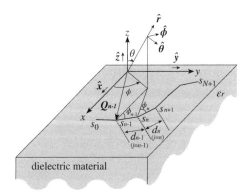

Figure 2.17   Coordinates for subsections on a dielectric half-space.

where

$$E_{s1}(s) = -\sum_{n=0}^{N} \int_{s_n}^{s_{n+1}} \frac{dI(s')}{ds'} \left[\frac{G^{(a)}(s+\Delta s^+, s')}{\Delta s} - \frac{G^{(a)}(s-\Delta s^-, s)}{\Delta s} \right] ds' \qquad (2.194)$$

$$E_{s2}(s) = k_0^2 \sum_{n=0}^{N} \int_{s_n}^{s_{n+1}} \Pi^{(a)s} I(s') \hat{s}' \bullet \hat{s} ds' \qquad (2.195)$$

with $G^{(a)}(s, s') = -\Pi^{(a)s}(s, s') + \Pi^{(a)}(s, s')$.

The MoM starts with expanding the unknown current $I(s')$ using a piecewise triangular function of

$$I(s') = \frac{s_{n+1} - s'}{d_n} I_n + \frac{s' - s_n}{d_n} I_{n+1} \quad s_n \leq s' \leq s_{n+1} \qquad (2.196)$$

Then, Eqs. (2.194) and (2.195) are

$$E_{s1}(s) = -\sum_{n=1}^{N} I_n \sum_{j=n-1}^{n} \eta_j \int_{s_j}^{s_{j+1}} \left[\frac{G^{(a)}(s+\Delta s^+, s')}{d_j \Delta s} - \frac{G^{(a)}(s-\Delta s^-, s')}{d_j \Delta s} \right] ds' \qquad (2.197)$$

$$E_{s2}(s) = k_0^2 \sum_{n=1}^{N} I_n \sum_{j=n-1}^{n} \int_{s_j}^{s_{j+1}} \frac{d_j - |s' - s_n|}{d_j} \times \Pi^{(a)s}(s, s') \hat{s}' \bullet \hat{s} ds' \qquad (2.198)$$

where $\eta_j = +1$ for $j = n-1$ and $\eta_j = -1$ for $j = n$.

Using Eqs. (2.197) and (2.198), Eq. (2.193) is written as

$$E_{in}(s) = \sum_{n=1}^{N} I_n \sum_{j=n-1}^{n} \int_{s_j}^{s_{j+1}} f_j(s, s') ds' \qquad (2.199)$$

where

$$f_j(s, s') = \frac{\eta_j}{d_j \Delta s} \left[G^{(a)}(s+\Delta s^+, s') - G^{(a)}(s-\Delta s^-, s') \right] - \frac{k_0^2}{d_j} (d_j - |s' - s_n|) \Pi^{(a)s}(s, s') \hat{s}' \bullet \hat{s} \qquad (2.200)$$

Equation (2.199) includes integral calculations. These are calculated adopting a trapezoidal formula. As a result, Eq. (2.199) is reduced to

$$E_{in}(s) = \sum_{n=1}^{N} I_n \sum_{j=n-1}^{n} \sum_{p=0}^{w_j} g(p) f_j(s, s_j + pT_j) T_j \qquad (2.201)$$

where $T_j = d_j/w_j$, $g(0) = g(w_j) = 0.5$ and $g(p) = 1$ for $p = 1, 2, \dots, w_j - 1$. The unknown expansion coefficient I_n is determined by solving the matrix expression $[Z][I] = [V]$ that is formed by applying the weighting functions to Eq. (2.201).

The use of the above-mentioned finite-difference technique has the advantage that the tangential electric field is expressed in a simple form, as shown in Eq. (2.201), where the need to integrate a poorly convergent first-order Bessel function, which derives from a derivative of $G^{(a)}(s, s')$ with respect to s, is obviated. Eq. (2.201) involves only single-integral calculations in $G^{(a)}(s, s') \left[= -\Pi^{(a)s}(s, s') + \Pi^{(a)}(s, s') \right]$ and $\Pi^{(a)s}(s, s')$.

Exercise [23]

Figure 2.18 shows the geometry where the space in $z > 0$ is filled with air and the space in $z < 0$ is filled with a homogeneous material characterized by permittivity ε_1 and permeability μ_1. The interface is denoted by i. The wavenumber (propagation phase constant) is denoted as $k_0 = \omega\sqrt{\varepsilon_0\mu_0}$ in the $z > 0$ space and $k_1 = \omega\sqrt{\varepsilon_1\mu_1}$ in the $z < 0$ space. Point P is an arbitrary field point (observation point) and point Q is a source point ($\mathbf{r}' = \mathbf{r}_s$), where an infinitesimal dipole source having the z-directed current moment Il_z is located.

The Hertzian vector $\mathbf{\Pi}^{(p)}$ as the primary field is expressed as

$$\Pi_z^{(p)}(r) = \frac{Il_z}{j\omega\varepsilon_0} G_0(\mathbf{r}, \mathbf{r}_s) = -\frac{Il_z}{4\pi\omega\varepsilon_0} \int_0^\infty \frac{\lambda}{h_0} J_0(\lambda|\boldsymbol{\rho} - \boldsymbol{\rho}_s|) e^{-jh_0|z - z_s|} d\lambda \qquad (2.202)$$

where $G_0(\mathbf{r}, \mathbf{r}_s) = e^{-jk_0|\mathbf{r} - \mathbf{r}_s|}/4\pi|\mathbf{r} - \mathbf{r}_s|$, $h_0 = \sqrt{k_0^2 - \lambda^2}$, $\mathbf{r}_s = \boldsymbol{\rho}_s + z_s\hat{z}$, and $J_0(..)$ is the Bessel function of the first kind of order 0. Formulate the scattered field in $z > 0$, $\Pi_z^{(s)}$, and the transmitted field in $z < 0$, $\Pi_z^{(t)}$.

Answer   The divergence of the Hertzian vector $\mathbf{\Pi} = \Pi_z\hat{z}$ is

$$\nabla \cdot (\Pi_z\hat{z}) = \frac{\partial \Pi_z}{\partial z} \qquad (2.203)$$

Hence,

$$\nabla\nabla \cdot (\Pi_z\hat{z}) = \frac{\partial^2 \Pi_z}{\partial\rho\partial z}\hat{\rho} + \frac{1}{\rho}\frac{\partial^2 \Pi_z}{\partial\phi\partial z}\hat{\phi} + \frac{\partial^2 \Pi_z}{\partial z^2}\hat{z} \qquad (2.204)$$

where a spherical coordinate system is used. The curl of $\mathbf{\Pi} = \Pi_z\hat{z}$ is

$$\nabla \times (\Pi_z\hat{z}) = \frac{1}{\rho}\frac{\partial \Pi_z}{\partial\phi}\hat{\rho} - \frac{\partial \Pi_z}{\partial\rho}\hat{\phi} \qquad (2.205)$$

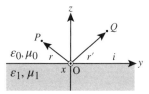

Figure 2.18   Current source above the interface.

Based on these results, the electromagnetic fields $E = \nabla\nabla \cdot \boldsymbol{\Pi} + k^2\boldsymbol{\Pi}$ and $H = j\omega\varepsilon\nabla \times \boldsymbol{\Pi}$ in a homogeneous medium characterized by ε, μ, and $k = \omega\sqrt{\varepsilon\mu}$ are written as

$$E = \frac{\partial^2\Pi_z}{\partial\rho\partial z}\hat{\rho} + \frac{1}{\rho}\frac{\partial^2\Pi_z}{\partial\phi\partial z}\hat{\phi} + \left(k^2\Pi_z + \frac{\partial^2\Pi_z}{\partial z^2}\right)\hat{z}, \qquad H = j\omega\varepsilon\left[\frac{1}{\rho}\frac{\partial\Pi_z}{\partial\phi}\hat{\rho} - \frac{\partial\Pi_z}{\partial\rho}\hat{\phi}\right] \qquad (2.206)$$

Let the electromagnetic fields above interface i and the corresponding Hertzian vector be $E^{(0)}$ and $H^{(0)}$ and $\Pi_z^{(0)}$, respectively, and let those below the interface be $E^{(1)}$ and $H^{(1)}$ and $\Pi_z^{(1)}$, respectively. Then, the boundary conditions are written as $\hat{z} \times E^{(0)} = \hat{z} \times E^{(1)}$ and $\hat{z} \times H^{(0)} = \hat{z} \times H^{(1)}$, since the unit vector normal to the interface is $\hat{z}$. Using Eq. (2.206), the boundary conditions are described as follows:

$$\begin{cases} \dfrac{\partial^2\Pi_z^{(0)}}{\partial\rho\partial z}\hat{\phi} - \dfrac{1}{\rho}\dfrac{\partial^2\Pi_z^{(0)}}{\partial\phi\partial z}\hat{\rho} = \dfrac{\partial^2\Pi_z^{(1)}}{\partial\rho\partial z}\hat{\phi} - \dfrac{1}{\rho}\dfrac{\partial^2\Pi_z^{(1)}}{\partial\phi\partial z}\hat{\rho} \\[3mm] \varepsilon_0\left[\dfrac{1}{\rho}\dfrac{\partial\Pi_z^{(0)}}{\partial\phi}\hat{\phi} + \dfrac{\partial\Pi_z^{(0)}}{\partial\rho}\hat{\rho}\right] = \varepsilon_1\left[\dfrac{1}{\rho}\dfrac{\partial\Pi_z^{(1)}}{\partial\phi}\hat{\phi} + \dfrac{\partial\Pi_z^{(1)}}{\partial\rho}\hat{\rho}\right] \end{cases} \qquad (2.207)$$

The above equations hold for all values of ρ and ϕ at $z = 0$, and hold for each component independently. Thus, the boundary conditions reduce to

$$\begin{cases} \dfrac{\partial\Pi_z^{(0)}}{\partial z} = \dfrac{\partial\Pi_z^{(1)}}{\partial z} \\[3mm] \varepsilon_0\Pi_z^{(0)} = \varepsilon_1\Pi_z^{(1)} \end{cases} \qquad (2.208)$$

where $\Pi_z^{(0)}$ is the sum of the contribution from the primary field and the scattered field, and $\Pi_z^{(1)}$ corresponds to the transmitted field, that is,

$$\Pi_z^{(0)} = \Pi_z^{(p)} + \Pi_z^{(s)} \qquad z \geq 0 \qquad (2.209a)$$

$$\Pi_z^{(1)} = \Pi_z^{(t)} \qquad z \leq 0 \qquad (2.209b)$$

From Eq. (2.202), the primary component of the Hertzian vector at the interface reduces to

$$\Pi_z^{(p)}(\rho, \phi, 0) = -\frac{Il_z}{4\pi\omega\varepsilon_0}\int_0^\infty \frac{\lambda}{h_0}J_0(\lambda|\boldsymbol{\rho} - \boldsymbol{\rho}_s|)e^{-jh_0 z_s}\mathrm{d}\lambda \qquad (2.210)$$

Since the boundary conditions of Eq. (2.208) with Eq. (2.209) are satisfied for all values of ρ and ϕ for an arbitrary value of $z_s > 0$, $J_0(\lambda|\boldsymbol{\rho} - \boldsymbol{\rho}_s|)e^{-jh_0 z_s}$ should be a common term in the integrand for $\Pi_z^{(s)}$ and $\Pi_z^{(t)}$.

The scattered field propagates in the $+z$ direction and the transmitted field propagates in the $-z$ direction. The wave moving in the $+z$ direction in the air can be expressed as $e^{-jh_0 z}$. Similarly, the wave moving in the $-z$ direction in the material can be expressed as $e^{jh_1 z}$. Hence, these terms are included in the integrands for $\Pi_z^{(s)}$ and $\Pi_z^{(t)}$ in the same manner as for $\Pi_z^{(p)}$. Thus, $\Pi_z^{(s)}$ and $\Pi_z^{(t)}$ are formulated as

$$\Pi_z^{(s)}(\boldsymbol{r}) = C\int_0^\infty \frac{\lambda}{h_0}R^{(\mathrm{TM})}(\lambda)J_0(\lambda|\boldsymbol{\rho} - \boldsymbol{\rho}_s|)e^{-jh_0(z+z_s)}\mathrm{d}\lambda \qquad (2.211a)$$

$$\Pi_z^{(t)}(\boldsymbol{r}) = C\int_0^\infty \frac{\lambda}{h_0}T^{(\mathrm{TM})}(\lambda)J_0(\lambda|\boldsymbol{\rho} - \boldsymbol{\rho}_s|)e^{j(h_1 z - h_0 z_s)}\mathrm{d}\lambda \qquad (2.211b)$$

where the constant C is $-\frac{Il_z}{4\pi\omega\varepsilon_0}$. $R^{(\mathrm{TM})}$ and $T^{(\mathrm{TM})}$ are constants to be determined.

From Eqs. (2.211a), (2.211b), and (2.208), the following equations are derived at $z = 0$

$$\begin{cases} h_0 \left[1 - R^{(TM)} \right] = h_1 T^{(TM)} \\ \varepsilon_0 \left[1 + R^{(TM)} \right] = \varepsilon_1 T^{(TM)} \end{cases} \tag{2.212}$$

Therefore, $R^{(TM)}$ and $T^{(TM)}$ are given by

$$\begin{cases} R^{(TM)} = \dfrac{\varepsilon_1 h_0 - \varepsilon_0 h_1}{\varepsilon_1 h_0 + \varepsilon_0 h_1} \\ T^{(TM)} = \dfrac{2\varepsilon_0 h_0}{\varepsilon_1 h_0 + \varepsilon_0 h_1} = \dfrac{\varepsilon_0}{\varepsilon_1} \left[1 + R^{(TM)} \right] \end{cases} \tag{2.213}$$

Thus, the required formulation of $\Pi_z^{(s)}$ and $\Pi_z^{(t)}$ is obtained.   ∎

REFERENCES

1. J. D. Jackson, *Classical Electrodynamics*, 3rd edn, New York: Wiley, 1999.
2. A. Arbabi and S. Safavi-Naeini, Maximum gain of a lossy antenna. *IEEE Trans. Antennas Propag.*, vol. 60, no. 1, pp. 2–7, 2012.
3. K. Fujita and H. Shirai, A study of the antenna radiation efficiency for electrically small antennas. In: *Proc. IEEE AP-S Int. Sympo.*, pp. 1522–1523, 2013.
4. H. Pocklington, Electrical oscillation in wires. *Camb. Philos. Soc. Proc.*, vol. 9, pp. 324–332, 1897.
5. E. Yamashita (ed), *Analysis Methods for Electromagnetic Wave Problems*, Vol. II, chapter 3, Boston: Artech House, 1996.
6. J. Kraus and R. Marhefka, *Antennas*, 3rd edn, New York: McGraw-Hill, 2002.
7. R. F. Harrington, *Field Computation by Moment Methods*, New York: Macmillan, 1968.
8. J. Moore and R. Pizer, *Moment Methods in Electromagnetics*, England: Research studies press, 1984.
9. K. Mei, On the integral equations of thin wire antennas. *IEEE Trans. Antennas Propag.*, vol. 13, pp. 374–378, 1965.
10. E. Hallen, Theoretical investigation into the transmitting and receiving qualities of antennae. *Nova Acta Regiae Soc. Sci. Upsal.*, vol. 11, no. 4, pp. 1–44, 1938.
11. H. Nakano, The simplified expression for the kernel of Mei's integral equation. *Trans. IECE Jpn.*, vol. J62-B, no. 11, pp. 1058–1059, 1979.
12. H. Nakano, The integral equations for a system composed of many arbitrarily bent wires. *Trans. IECE Jpn.*, vol. E65, no. 6, pp. 303–309, 1982.
13. H. Nakano, et al., Integral equations on electromagnetic coupling to wires through apertures, IECE TGAP, Part 1, AP-82-49, 1982; Part 2, AP-83-59, 1983; Part 3, AP-83-60, 1983; Part 4, AP-84-38, 1984; and Part 5, AP-84-113, 1985.
14. H. Nakano and R. Harrington, Integral equation on electromagnetic coupling to a wire through an aperture. *Trans. IECE Jpn.*, Vol. E66, No. 6, 1983, pp. 383–389.
15. I. Rana and N. Alexopoulos, Current distribution and input impedance of printed dipoles. *IEEE Trans. Antennas Propag.*, vol. 29, no. 1, pp. 99–105, 1981.
16. H. Nakano, S. Kerner, and N. Alexopoulos, The moment method solution for printed wire antennas of arbitrary configuration. *IEEE Trans. Antennas Propag.*, vol. 36, no. 12, pp. 1667–1674, 1988.
17. H. Nakano, K. Hirose, T. Suzuki, S. R. Kerner, and N.G. Alexopoulos, Numerical analyses of printed line antennas. *IEE Proc. Microw. Antennas Propag. H*, vol. 136, no. 2, pp. 98–104, 1989.
18. H. Nakano, A numerical approach to line antennas printed on dielectric materials. *Comput. Phys. Commun.*, vol. 68, pp. 441–450, 1991.
19. H. Nakano, T. Kawano, Y. Kozono, and J. Yamauchi, A fast MoM calculation technique using sinusoidal basis and testing functions for a wire on a dielectric substrate and its application to meander loop and grid array antennas. *IEEE Trans. Antennas Propag.*, vol. AP-53, no. 10, pp. 3300–3307, 2005.

20. D. R. Jackson and N. G. Alexopoulos, An asymptotic extraction technique for evaluating Sommerfeld-type integrals. *IEEE Trans. Antennas Propag.*, vol. 34, no. 12, pp. 1467–1470, 1986.
21. H. Nakano, K. Hirose, I. Ohshima, and J. Yamauchi, An integral equation and its application to spiral antennas on semi-infinite dielectric materials. *IEEE Trans. Antennas Propag.*, vol. AP-46, no. 2, pp. 267–274, 1998.
22. H. Nakano, K. Hirose, M. Yamazaki, and J. Yamauchi, Square spiral antenna on dielectric half-space: analysis using an electric field equation formulated by a finite-difference techniques. *IEE Proc. Microw. Antennas Propag.*, vol. 145, no. 1, pp. 70–74, 1998.
23. T. Uno, Private communication, January 2015.
24. K. Fujita, Private communication, January 2015.

Chapter 3

Finite-Difference Time-Domain Methods (FDTDMs)

3.1 BASIS

The current distributed along the antenna in Sections 2.2–2.9 is used for calculating the antenna characteristics. The finite-difference time-domain method (FDTDM) in this section is the method to obtain the electric and magnetic fields for a space that includes the antenna [1,2]. These fields are in turn used to calculate the antenna characteristics, as will be seen later.

Figure 3.1a shows an FDTD analysis space of volume $X \times Y \times Z$. This analysis space is subdivided into small parallelepipeds (cells), each having a volume of $\Delta x \times \Delta y \times \Delta z$. For applying Maxwell's equations to the analysis space, field components (E_x, E_y, E_z) and (H_x, H_y, H_z) are assigned to each cell, as shown in Fig. 3.1b, and the spatial coordinates (x, y, z) and time coordinate t are expressed as $(i\Delta x, j\Delta y, k\Delta z)$ and $n\Delta t$, respectively, with i, j, k, and t being integers. Note that absorbing boundary conditions (ABC) [3–7] are used to terminate the outer surface of this analysis space.

For the FDTDM, the function $g(x, y, z, t) = g(i\Delta x, j\Delta y, k\Delta z, n\Delta t)$ is expressed as $g^n(i, j, k)$, and the derivative of this function with respect to the spatial coordinates, for instance, the coordinate x, is expressed using a central finite difference as

$$\frac{\partial g^n(i,j,k)}{\partial x} \approx \frac{g^n(i+(1/2),j,k) - g^n(i-(1/2),j,k)}{\Delta x} \tag{3.1}$$

In addition, the derivatives of the electric and magnetic fields with respect to time t are expressed as

$$\left. \frac{\partial \mathbf{E}}{\partial t} \right|_{t=(n-(1/2))\Delta t} \approx \frac{\mathbf{E}^n - \mathbf{E}^{n-1}}{\Delta t} \tag{3.2}$$

$$\left. \frac{\partial \mathbf{H}}{\partial t} \right|_{t=n\Delta t} \approx \frac{\mathbf{H}^{n+(1/2)} - \mathbf{H}^{n-(1/2)}}{\Delta t} \tag{3.3}$$

Low-Profile Natural and Metamaterial Antennas: Analysis Methods and Applications, First Edition.
Hisamatsu Nakano.
© 2016 The Institute of Electrical and Electronics Engineers, Inc. Published 2016 by John Wiley & Sons, Inc.

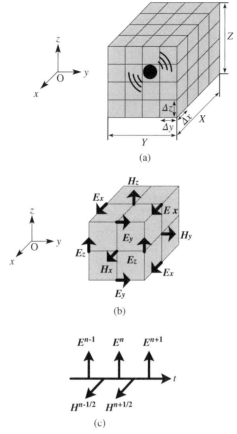

Figure 3.1   FDTD analysis space. (a) Subdivisions (cells). (b) Assignment of field components to a cell. (c) Assignment of time.

where superscripts $(n - 1, n)$ and $(n - 1/2, n + 1/2)$ denote the time coordinates assigned to the electric and magnetic fields, respectively (see Fig. 3.1c). Furthermore, Maxwell's curl equations for the electric field at time t and the magnetic field at time $t - 1/2$

$$\nabla \times \mathbf{E}|_{t=n\Delta t} = -\mu \frac{\partial \mathbf{H}}{\partial t}\bigg|_{t=n\Delta t} \tag{3.4}$$

$$\nabla \times \mathbf{H}|_{t=(n-(1/2))\Delta t} = \sigma \mathbf{E}|_{t=(n-(1/2))\Delta t} + \varepsilon \frac{\partial \mathbf{E}}{\partial t}\bigg|_{t=(n-(1/2))\Delta t} \tag{3.5}$$

are expressed as

$$\mathbf{H}^{n+(1/2)} = \mathbf{H}^{n-(1/2)} - \frac{\Delta t}{\mu} \nabla \times \mathbf{E}^n \tag{3.6}$$

$$\mathbf{E}^n = \frac{2\varepsilon - \sigma\Delta t}{2\varepsilon + \sigma\Delta t} \mathbf{E}^{n-1} + \frac{2\Delta t}{2\varepsilon + \sigma\Delta t} \nabla \times \mathbf{H}^{n-(1/2)} \tag{3.7}$$

where ε, μ, and σ are the permittivity, permeability, and conductivity, respectively. Note that $\sigma\mathbf{E}$ in Eq. (3.5) is approximated by the average of $\sigma\mathbf{E}^n$ and $\sigma\mathbf{E}^{n-1}$ to obtain Eq. (3.7), because the electric field at $t = (n - (1/2))\Delta t$ is not assigned to the time coordinate.

Using Eq. (3.1), the finite-difference form for the x-component of Eq. (3.7) is written as

$$
\begin{aligned}
E_x^n(i + (1/2), j, k) &= e_{x/0}(i + (1/2), j, k)E_x^{n-1}(i + (1/2), j, k) \\
&+ e_{x/y}(i + (1/2), j, k)[H_z^{n-(1/2)}(i + (1/2), j + (1/2), k) - H_z^{n-(1/2)}(i + (1/2), j - (1/2), k)] \\
&- e_{x/z}(i + (1/2), j, k)[H_y^{n-(1/2)}(i + (1/2), j, k + (1/2)) - H_y^{n-(1/2)}(i + (1/2), j, k - (1/2))]
\end{aligned}
$$

$$(3.8)$$

where

$$
e_{x/0}(i + (1/2), j, k) = \frac{2\varepsilon(i + (1/2), j, k) - \sigma(i + (1/2), j, k)\Delta t}{2\varepsilon(i + (1/2), j, k) + \sigma(i + (1/2), j, k)\Delta t}
$$

$$(3.9)$$

$$
e_{x/y}(i + (1/2), j, k) = \frac{2\Delta t}{2\varepsilon(i + (1/2), j, k) + \sigma(i + (1/2), j, k)\Delta t}\frac{1}{\Delta y}
$$

$$(3.10)$$

$$
e_{x/z}(i + (1/2), j, k) = \frac{2\Delta t}{2\varepsilon(i + (1/2), j, k) + \sigma(i + (1/2), j, k)\Delta t}\frac{1}{\Delta z}
$$

$$(3.11)$$

The two other components of electric field $\mathbf{E}^n$ and the three components of magnetic field $\mathbf{H}^{n+(1/2)}$ are similarly formulated.

The solutions to Maxwell's equations, $\mathbf{E}$ and $\mathbf{H}$, are obtained by iterating Eqs. (3.6) and (3.7) with $n = 1, 2, \ldots$ until these fields become constant (converge). This iterative method is called the finite-difference time-domain method (FDTDM). The following [I]–[III] describe the postprocesses for obtaining the antenna characteristics.

[I]: the θ and ϕ components of the radiation field (in the frequency domain) are revealed by applying the equivalence theorem [2] to the obtained FDTDM $\mathbf{E}$ and $\mathbf{H}$ fields;

$$
E_\theta(\omega) = -\frac{jk_0}{4\pi}\frac{e^{-jk_0 r}}{r}\left[Z_0\mathbf{N}(\omega) \bullet \hat{\theta} + \mathbf{L}(\omega) \bullet \hat{\phi}\right]
$$

$$(3.12)$$

$$
E_\phi(\omega) = -\frac{jk_0}{4\pi}\frac{e^{-jk_0 r}}{r}\left[Z_0\mathbf{N}(\omega) \bullet \hat{\phi} - \mathbf{L}(\omega) \bullet \hat{\theta}\right]
$$

$$(3.13)$$

where the spherical coordinates (r, θ, ϕ) are used with unit vectors $(\hat{r}, \hat{\theta}, \hat{\phi})$. Z_0 is the intrinsic impedance ($120\pi\ \Omega$), k_0 is the propagation phase constant in free space, and

$$
\mathbf{N}(\omega) = \int_{\text{closed surface}} \mathbf{J}_s(\omega, \mathbf{r}')e^{jk_0 \hat{r} \bullet \mathbf{r}'}\,dS'
$$

$$(3.14)$$

$$
\mathbf{L}(\omega) = \int_{\text{closed surface}} \mathbf{M}_s(\omega, \mathbf{r}')e^{jk_0 \hat{r} \bullet \mathbf{r}'}\,dS'
$$

$$(3.15)$$

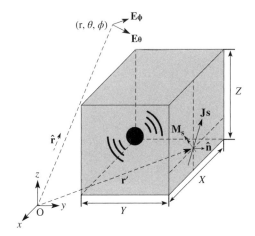

Figure 3.2   Equivalence theorem.

The $\mathbf{J}_s(\omega, \mathbf{r}')$ and $\mathbf{M}_s(\omega, \mathbf{r}')$ in Eqs. (3.14) and (3.15) are, respectively, the Fourier transform of the time-domain electric current density $\mathbf{J}_s(t, \mathbf{r}') = \hat{n} \times \mathbf{H}(t, \mathbf{r}')$ and of the magnetic current density $\mathbf{M}_s(t, \mathbf{r}') = \mathbf{E}(t, \mathbf{r}') \times \hat{n}$, where $\mathbf{r}'$ is the position vector from the coordinate origin to a point located on the closed surface of the analysis space, and $\hat{n}$ is the outward unit vector normal to the closed surface (see Fig. 3.2).

[II]: by integrating the obtained FDTDM magnetic field around the antenna conductor, the current along the antenna conductor, $I(t)$, is revealed (Ampere's law); by taking the product of the FDTDM electric field and the distance between two points, the voltage between the two points, $V(t)$, is revealed. Based on these results, the input impedance $Z_{in} = R_{in} + jX_{in}$ is calculated using $F[V_{in}(t)]/F[I_{in}(t)]$, where $F[V_{in}(t)]$ and $F[I_{in}(t)]$ are the Fourier-transformed antenna input voltage and current, respectively. [III]: the gain is obtained by using the revealed radiation field and the calculated input impedance.

3.2 LOD–FDTD METHOD

The FDTDM has the limitation of having a time step caused by the Courant–Friedrichs–Lewy (CFL) condition [2]. In particular, the time step becomes extremely small when spatial sampling meshes are chosen to be small. This situation gives rise to a long computation time. To alleviate this problem, implicit FDTDMs have been formulated with the help of the alternating direction implicit (ADI) method and the Crank–Nicolson (CN) method [8–10]. For practical simulations, the implicit FDTDMs can be efficient, owing to the use of a large time step that is beyond the CFL limit.

Recently, the locally one-dimensional (LOD) method has been applied to the implicit FDTDM [11,12] and improved [13–18]. The advantage of the

LOD–FDTDM is a simpler implementation of the algorithm, in comparison with the ADI and CN counterparts. In the following, the LOD–FDTDM is presented and discussed.

Maxwell's equations for isotropic, linear, and lossless materials are expressed as

$$\frac{\partial u}{\partial t} = ([A] + [B])\vec{u} \tag{3.16}$$

where

$$\vec{u} = \begin{bmatrix} E_x & E_y & E_z & H_x & H_y & H_z \end{bmatrix}^T \tag{3.17}$$

$$[A] = \begin{bmatrix} 0 & 0 & 0 & 0 & -\dfrac{\gamma c}{\varepsilon_r}\dfrac{\partial}{\partial z} & 0 \\ 0 & 0 & 0 & 0 & 0 & -\dfrac{\alpha c}{\varepsilon_r}\dfrac{\partial}{\partial x} \\ 0 & 0 & 0 & -\dfrac{\beta c}{\varepsilon_r}\dfrac{\partial}{\partial y} & 0 & 0 \\ 0 & 0 & -\dfrac{\beta c}{\mu_r}\dfrac{\partial}{\partial y} & 0 & 0 & 0 \\ -\dfrac{\gamma c}{\mu_r}\dfrac{\partial}{\partial z} & 0 & 0 & 0 & 0 & 0 \\ 0 & -\dfrac{\alpha c}{\mu_r}\dfrac{\partial}{\partial x} & 0 & 0 & 0 & 0 \end{bmatrix} \tag{3.18}$$

$$[B] = \begin{bmatrix} 0 & 0 & 0 & 0 & 0 & \dfrac{\beta c}{\varepsilon_r}\dfrac{\partial}{\partial y} \\ 0 & 0 & 0 & \dfrac{\gamma c}{\varepsilon_r}\dfrac{\partial}{\partial z} & 0 & 0 \\ 0 & 0 & 0 & 0 & \dfrac{\alpha c}{\varepsilon_r}\dfrac{\partial}{\partial x} & 0 \\ 0 & \dfrac{\gamma c}{\mu_r}\dfrac{\partial}{\partial z} & 0 & 0 & 0 & 0 \\ 0 & 0 & \dfrac{\alpha c}{\mu_r}\dfrac{\partial}{\partial x} & 0 & 0 & 0 \\ \dfrac{\beta c}{\mu_r}\dfrac{\partial}{\partial y} & 0 & 0 & 0 & 0 & 0 \end{bmatrix} \tag{3.19}$$

in which c is the speed of light in a vacuum, ε_r is the relative permittivity, and μ_r is the relative permeability. To equalize the field amplitudes between E and H, the normalized expression of the fields is employed. α, β, and γ, respectively, represent the dispersion control parameters in the x, y, and z directions, and will be discussed later.

Discretizing Eq. (3.16) with the central difference results in the traditional explicit FDTDM, where the time step is constrained by the CFL condition. To remove this constraint, the Crank–Nicolson (CN) scheme [10] is applied to

Eq. (3.16), leading to

$$\vec{u}^{n+1} = \frac{[I] + (\Delta t/2)([A] + [B])}{[I] - (\Delta t/2)([A] + [B])} \vec{u}^n \tag{3.20}$$

where $[I]$ is the 6×6 identity matrix. Factoring out Eq. (3.20), assuming that $(\Delta t^2/4)[A][B]$ is negligibly small, gives

$$\vec{u}^{n+1} = \frac{\left([I] + \frac{\Delta t}{2}[A]\right)\left([I] + \frac{\Delta t}{2}[B]\right)}{\left([I] - \frac{\Delta t}{2}[A]\right)\left([I] - \frac{\Delta t}{2}[B]\right)} \vec{u}^n \tag{3.21}$$

The above equation is solved with the LOD method, using the intermediate field term $\vec{u}^{n+(1/2)}$

$$\vec{u}^{n+(1/2)} = \frac{([I] + (\Delta t/2)[B])}{([I] - (\Delta t/2)[B])} \vec{u}^n \tag{3.22}$$

$$\vec{u}^{n+1} = \frac{([I] + (\Delta t/2)[A])}{([I] - (\Delta t/2)[A])} \vec{u}^{n+(1/2)} \tag{3.23}$$

which are formulated to be Eqs. (3.24)–(3.29) and (3.30)–(3.35), respectively.

$$E_x^{n+(1/2)} = E_x^n + \frac{\beta c \Delta t}{2\varepsilon_r}\left(\frac{\partial H_z^{n+(1/2)}}{\partial y} + \frac{\partial H_z^n}{\partial y}\right) \tag{3.24}$$

$$E_y^{n+(1/2)} = E_y^n + \frac{\gamma c \Delta t}{2\varepsilon_r}\left(\frac{\partial H_x^{n+(1/2)}}{\partial z} + \frac{\partial H_x^n}{\partial z}\right) \tag{3.25}$$

$$E_z^{n+(1/2)} = E_z^n + \frac{\alpha c \Delta t}{2\varepsilon_r}\left(\frac{\partial H_y^{n+(1/2)}}{\partial x} + \frac{\partial H_y^n}{\partial x}\right) \tag{3.26}$$

$$H_x^{n+(1/2)} = H_x^n + \frac{\gamma c \Delta t}{2\mu_r}\left(\frac{\partial E_y^{n+(1/2)}}{\partial z} + \frac{\partial E_y^n}{\partial z}\right) \tag{3.27}$$

$$H_y^{n+(1/2)} = H_y^n + \frac{\alpha c \Delta t}{2\mu_r}\left(\frac{\partial E_z^{n+(1/2)}}{\partial x} + \frac{\partial E_z^n}{\partial x}\right) \tag{3.28}$$

$$H_z^{n+(1/2)} = H_z^n + \frac{\beta c \Delta t}{2\mu_r}\left(\frac{\partial E_x^{n+(1/2)}}{\partial y} + \frac{\partial E_x^n}{\partial y}\right) \tag{3.29}$$

$$E_x^{n+1} = E_x^{n+(1/2)} - \frac{\gamma c \Delta t}{2\varepsilon_r}\left(\frac{\partial H_y^{n+1}}{\partial z} + \frac{\partial H_y^{n+(1/2)}}{\partial z}\right) \tag{3.30}$$

$$E_y^{n+1} = E_y^{n+(1/2)} - \frac{\alpha c \Delta t}{2\varepsilon_r}\left(\frac{\partial H_z^{n+1}}{\partial x} + \frac{\partial H_z^{n+(1/2)}}{\partial x}\right) \tag{3.31}$$

$$E_z^{n+1} = E_z^{n+(1/2)} - \frac{\beta c \Delta t}{2\varepsilon_r} \left(\frac{\partial H_x^{n+1}}{\partial y} + \frac{\partial H_x^{n+(1/2)}}{\partial y} \right) \tag{3.32}$$

$$H_x^{n+1} = H_x^{n+(1/2)} - \frac{\beta c \Delta t}{2\mu_r} \left(\frac{\partial E_z^{n+1}}{\partial y} + \frac{\partial E_z^{n+(1/2)}}{\partial y} \right) \tag{3.33}$$

$$H_y^{n+1} = H_y^{n+(1/2)} - \frac{\gamma c \Delta t}{2\mu_r} \left(\frac{\partial E_x^{n+1}}{\partial z} + \frac{\partial E_x^{n+(1/2)}}{\partial z} \right) \tag{3.34}$$

$$H_z^{n+1} = H_z^{n+(1/2)} - \frac{\alpha c \Delta t}{2\mu_r} \left(\frac{\partial E_y^{n+1}}{\partial x} + \frac{\partial E_y^{n+(1/2)}}{\partial x} \right) \tag{3.35}$$

Eqs. (3.29), (3.27), and (3.28) are substituted into Eqs. (3.24), (3.25), and (3.26), respectively, and the resultant equations are solved implicitly. Then, Eqs. (3.29), (3.27), and (3.28) are solved explicitly. Similarly, Eqs. (3.30)–(3.35) are solved. Note that each equation to be solved contains an x, y, or z derivative leading to a locally one-dimensional calculation.

Next, the dispersion control parameters α, β, and γ are derived on the basis of the numerical dispersion analysis. The numerical dispersion relation is obtained by substituting a plane wave expressed by $e^{j(\omega p \Delta t - k_x q \Delta x - k_y l \Delta y - k_z m \Delta z)}$ into the FD equations, where p, q, l, and m denote the indexes for the t, x, y, and z axes, respectively, ω ($= kc$ with k being the free-space wavenumber) is the angular frequency, and k_x, k_y, and k_z are the wavenumbers in the x, y, and z directions, respectively. When the wave propagation is assumed to be in one direction, the dispersion relation, for example, in the x direction, is derived as

$$\tan \frac{\omega \Delta t}{2} = \alpha \left(\frac{c \Delta t}{\underline{n} \Delta x} \right) \sin \frac{k_x \Delta x}{2} \tag{3.36}$$

where $\underline{n} = \sqrt{\varepsilon_r}$. As a result, the dispersion control parameter is given by Eq. (3.37). Similarly, β and γ are obtained as Eqs. (3.38) and (3.39), respectively:

$$\alpha = \left(\frac{\underline{n} \Delta x}{c \Delta t} \right) \frac{\tan(\omega \Delta t / 2)}{\sin(k_x \underline{n} \Delta x / 2)} \tag{3.37}$$

$$\beta = \left(\frac{\underline{n} \Delta y}{c \Delta t} \right) \frac{\tan(\omega \Delta t / 2)}{\sin\left(k_y \underline{n} \Delta y / 2\right)} \tag{3.38}$$

$$\gamma = \left(\frac{\underline{n} \Delta z}{c \Delta t} \right) \frac{\tan(\omega \Delta t / 2)}{\sin(k_z \underline{n} \Delta z / 2)} \tag{3.39}$$

For sufficiently small spatial sampling widths that provide converged solutions, $\sin\theta$ is approximated to θ, so the above expressions are simplified to

$$\alpha = \frac{2 \tan(\omega \Delta t / 2)}{c \Delta t k_x} \tag{3.40}$$

$$\beta = \frac{2\tan(\omega \Delta t/2)}{c \Delta t k_y} \tag{3.41}$$

$$\gamma = \frac{2\tan(\omega \Delta t/2)}{c \Delta t k_z} \tag{3.42}$$

Note that knowledge regarding the index $\underline{n}$ is not required in the numerical dispersion parameters α, β, and γ. In other words, the control parameters can be used without paying attention to the index n, provided that the spatial sampling width is sufficiently small. Numerical examples are found in Refs 19 and 20.

Exercise

Derive Eq. (3.37).

Answer   We focus on the wave propagation in the x-direction. Eqs. (3.32) and (3.34) for this case are written as

$$E_z^{n+1} = E_z^{n+(1/2)} \tag{3.43}$$

$$H_y^{n+1} = H_y^{n+(1/2)} \tag{3.44}$$

Substituting Eqs. (3.43) and (3.44) into Eqs. (3.26) and (3.28) leads to

$$E_z^{n+1} - E_z^n = \frac{\alpha c \Delta t}{2\varepsilon_r} \left(\frac{\partial H_y^{n+1}}{\partial x} + \frac{\partial H_y^n}{\partial x} \right) \tag{3.45}$$

$$H_y^{n+1} - H_y^n = \frac{\alpha c \Delta t}{2\mu_r} \left(\frac{\partial E_z^{n+1}}{\partial x} + \frac{\partial E_z^n}{\partial x} \right) \tag{3.46}$$

The plane wave propagating in the x-direction is expressed as

$$\begin{bmatrix} E_{z0} \\ H_{y0} \end{bmatrix} e^{j(\omega p \Delta t - k_x q \Delta x)} \tag{3.47}$$

Applying Eq. (3.47) to Eq. (3.45) results in

$$E_{z0}e^{j[\omega(n+1)\Delta t - k_x q \Delta x]} - E_{z0}e^{j(\omega n \Delta t - k_x q \Delta x)}$$
$$= \frac{\alpha c \Delta t}{2\Delta x \varepsilon_r} \left[\begin{array}{l} \left(H_{y0}e^{j[\omega(n+1)\Delta t - k_x(q+(1/2))\Delta x]} - H_{y0}e^{j[\omega(n+1)\Delta t - k_x(q-(1/2))\Delta x]} \right) \\ + \left(H_{y0}e^{j[\omega n \Delta t - k_x(q+(1/2))\Delta x]} - H_{y0}e^{j[\omega n \Delta t - k_x(q-(1/2))\Delta x]} \right) \end{array} \right]$$

which is rewritten as

$$E_{z0}\left(e^{j\omega \Delta t} - 1\right) = \frac{\alpha c \Delta t}{2\Delta x \varepsilon_r} H_{y0}\left(e^{j\omega \Delta t} + 1\right) 2j \sin\frac{k_x \Delta x}{2} \tag{3.48}$$

Similarly, substituting Eq. (3.47) into Eq. (3.46) provides

$$H_{y0}\left(e^{j\omega \Delta t} - 1\right) = \frac{\alpha c \Delta t}{2\Delta x \mu_r} E_{z0}\left(e^{j\omega \Delta t} + 1\right) 2j \sin\frac{k_x \Delta x}{2} \tag{3.49}$$

Substituting H_{y0} of Eq. (3.49) into Eq. (3.48) with $\mu_r = 1$ gives the following relation:

$$\tan\frac{\omega\Delta t}{2} = \alpha\frac{c\Delta t}{\underline{n}\Delta x}\sin\frac{k_x\Delta x}{2} \tag{3.50}$$

From Eq. (3.50), the dispersion control parameter (3.37) is obtained.   ■

REFERENCES

1. K. S. Yee, Numerical solution of initial boundary value problems involving Maxwell's equations in isotropic media. *IEEE Trans. Antennas Propag.*, vol. 14, no. 3, pp. 302–307, 1966.
2. A. Taflove and S. C. Hagness, Computational electrodynamics. *The Finite-Difference Time-Domain Method*, 3rd edn, Norwood, MA: Artech House, 2005.
3. G. Mur, Absorbing boundary conditions for the finite-difference approximation of the time-domain electromagnetic field equation. *IEEE Trans. Electromagn. Compat.*, vol. EMC-23 no. 4, pp. 377–382, 1981.
4. Z. Liao, H. Wong, B. Yang and Y. Yuan, A transmitting boundary for transient wave analyses. *Sci. Sin. A*, vol. 27, No. 10, pp. 1063–1076, 1984.
5. R. Higdon, Absorbing boundary conditions for difference approximations to the multi-dimensional wave equation. *Math. Comput.*, vol. 47, no. 176, pp. 437–459, 1986.
6. R. Higdon, Numerical absorbing boundary conditions for the wave equation. *Math. Comput.*, vol. 49, no. 179, pp. 65–90, 1987.
7. J. Berenger, A perfectly matched layer for the absorption of electromagnetic waves. *J. Comput. Phys.*, vol. 114, no. 1, pp. 185–200, 1994.
8. T. Namiki, A new FDTD algorithm based on alternating-direction implicit method. *IEEE Trans. Microw. Theory Tech.*, vol. 47, no. 10, pp. 2003–2007, 1999.
9. F. H. Zheng, Z. Z. Chen, and J. Z. Zhang, A finite-difference time-domain method without the Courant stability conditions. *IEEE Microw. Guided Wave Lett.*, vol. 9, no. 11, pp. 441–443, 1999.
10. G. Sun and C.W. Trueman, Analysis and numerical experiments on the numerical dispersion of two-dimensional ADI-FDTD. *IEEE Antennas Wirel. Propagat. Lett.*, vol. 2, pp. 78–81, 2003.
11. J. Shibayama, M. Muraki, J. Yamauchi, and H. Nakano, Efficient implicit FDTD algorithm based on locally one-dimensional scheme. *Electron. Lett.*, vol. 41, no. 19, pp. 1046–1047, 2005.
12. V. E. do Nascimento, J. A. Cuminato, F. L. Teixeira, and B.-H. V. Borges, Unconditionally stable finite-difference time-domain method based on the locally-one-dimensional technique. In: *Proc. 22nd Symp. Brasileiro Telecomun.*, Campinas, Brazil, Sep. 2005, pp. 288–291.
13. E. L. Tan, Unconditionally stable LOD-FDTD method for 3-D Maxwell's equations. *IEEE Microw. Wirel. Compon. Lett.*, vol. 17, no. 2, pp. 85–87, 2007.
14. E. L. Tan, Fundamental schemes for efficient unconditionally stable implicit finite-difference time-domain methods. *IEEE Trans. Antennas Propagat.*, vol. 56, no. 1, pp. 170–177, 2008.
15. E. Li, I. Ahmed, and R. Vahldieck, Numerical dispersion analysis with an improved LOD-FDTD method. *IEEE Microw. Wirel. Compon. Lett.*, vol. 17, no. 5, pp. 319–321, 2007.
16. I. Ahmed, E. Li, and K. Krohne, Convolutional perfectly matched layer for an unconditionally stable LOD-FDTD method. *IEEE Microw. Wireless Compon. Lett.*, vol. 17, no. 12, pp. 816–818, 2007.
17. K.-Y. Jung and F. L. Teixeira, An iterative unconditionally stable LOD-FDTD method. *IEEE Microw. Wirel. Compon. Lett.*, vol. 18, no. 2, pp. 76–78, 2008.
18. O. Ramadan, Unsplit field implicit PML algorithm for complex envelope dispersive LOD-FDTD simulations. *Electron. Lett.*, vol. 43, no. 5, pp. 267–268, 2007.
19. J. Shibayama, R. Ando, J. Yamauchi, and H. Nakano, An LOD-FDTD method for the analysis of periodic structures at normal incidence. *IEEE Antennas Wirel. Propag. Lett.*, vol. 8, pp. 890–893, 2009.
20. J. Shibayama, R. Ando, J. Yamauchi, and H. Nakano, A 3-D LOD-FDTD method for the wideband analysis of optical devices. *J. Lightw. Technol.*, vol. 29, no. 11, pp. 1652–1658, 2011.

Part II

Low-Profile Natural Antennas

Part II-1

Base Station Antennas

Chapter 4

Inverted-F Antennas

In most cases, the height of a monopole is chosen to be approximately one-quarter wavelength above a ground plane (GP). Applications using a monopole often require an antenna height that is shorter than one-quarter wavelength. For such applications, the monopole is bent; this antenna is called the inverted-L antenna (ILA). A derivative of the ILA is called the inverted-F antenna (IFA) [1–3]. The IFA has the advantage that impedance matching to a feed line is easier than for the ILA. As a result, the VSWR frequency band for the IFA is moderately wide; the frequency bandwidth for a VSWR $= 2$ criterion is approximately 8% for an antenna height of approximately one-tenth wavelength ($\lambda/10$). This chapter presents a technique to further increase the VSWR bandwidth and discusses the radiation characteristics of the IFA.

4.1 INVERTED-F ANTENNA WITH A SINGLE PARASITIC INVERTED-L ELEMENT

Figure 4.1 shows an IFA with a parasitic inverted-L (IL) element [4–6] (separated by distance Δs from the IFA). The IFA and IL are made of wires of the same diameter $2a$ and are chosen to have the same height ($H_{IF} = H_{IL}$). The horizontal lengths for the IFA and IL are, respectively, L_{IF} and L_{IL} ($L_{IF} < L_{IL}$). The feed point for the IFA, F, is located at $(x, y) = (0, 0)$, and the coordinates for the bottom end of the z-directed conducting wire ab constituting the IFA are $(x, y) = (L_{FD}, 0)$, where L_{FD} is called the IFA feed distance. In the following discussion, the IL horizontal length L_{IL} and the IFA feed distance L_{FD} are changed according to the objectives of the discussion; the other parameters are fixed and are shown in Table 4.1. Note that the ground plane is assumed to be of infinite extent ($GP_x = GP_y = \infty$).

First, we change only the IL horizontal length L_{IL}, while holding the IFA feed distance L_{FD} fixed. Figure 4.2 shows the input impedance ($Z_{in} = R_{in} + jX_{in}$), where parameters ($L_{IL}$, L_{FD}) = (varied, 6.75 mm) = (varied, $0.027\lambda_{1.2}$) are used, together with the parameters in Table 4.1, where $\lambda_{1.2}$ is the wavelength at 1.2 GHz. Note

Low-Profile Natural and Metamaterial Antennas: Analysis Methods and Applications, First Edition.
Hisamatsu Nakano.
© 2016 The Institute of Electrical and Electronics Engineers, Inc. Published 2016 by John Wiley & Sons, Inc.

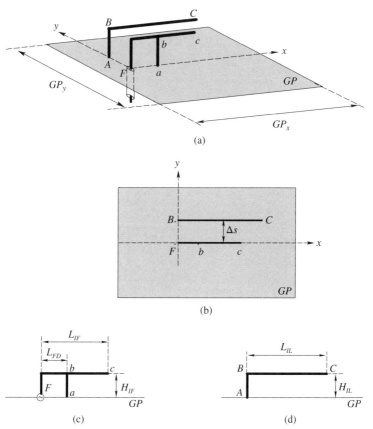

Figure 4.1   Inverted-F antenna (IFA) with a parasitic inverted-L element. (a) Perspective view. (b) Top view. (c) Side view of the fed inverted F. (d) Side view of the parasitic inverted L.

that this input impedance is obtained using integral equation $N2$. It is revealed that, as the IL horizontal length L_{IL} is increased, the lower resonance frequency f_L decreases ($X_{in} = 0$ at f_L), while the higher resonance frequency f_H is less influenced. An important observation is that the IL element length ABC ($= L_{IL} + H_{IL}$) at f_L corresponds to approximately one-quarter wavelength.

Table 4.1   Fixed Parameters

Symbol	Value	Symbol	Value
H_{IF}	22.75 mm $= 0.091\lambda_{1.2}$	Δs	6 mm $= 0.024\lambda_{1.2}$
L_{IF}	44 mm $= 0.176\lambda_{1.2}$	GP_x	∞
H_{IL}	22.75 mm $= 0.091\lambda_{1.2}$	GP_y	∞
$2a$	3 mm $= 0.012\lambda_{1.2}$	$\lambda_{1.2}$	250 mm

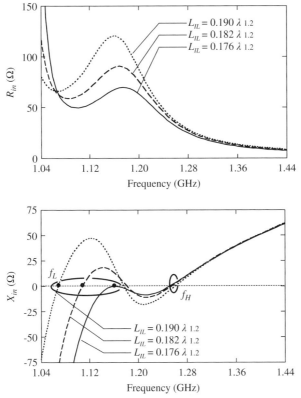

Figure 4.2 Input impedance $Z_{in} = R_{in} + jX_{in}$ for the IFA with a parasitic IL element, with the IL horizontal length L_{IL} as a parameter. Parameters $(L_{IL}, L_{FD}) = $ (varied, 6.75 mm) = (varied, $0.027\lambda_{1.2}$) and those shown in Table 4.1 are used. (Reproduced from Ref. [6] with permission from IET.)

Next, we change only the IFA feed distance L_{FD}, while holding the IL horizontal length L_{IL} fixed. Parameters $(L_{IL}, L_{FD}) = $ (45.5 mm, varied) = ($0.182\lambda_{1.2}$, varied) and those shown in Table 4.1 are used. Figure 4.3 shows the frequency response of the input impedance. It is found that a change in the IFA feed distance L_{FD} remarkably influences the higher resonance frequency f_H, with less effect on the lower resonance frequency f_L. One finding is that the length abc for the IFA [$= (L_{IF} - L_{FD}) + H_{IF}$] at f_H corresponds to approximately one-quarter wavelength.

The phenomena shown in Figs. 4.2 and 4.3 indicate that the lower and higher resonance frequencies are independently controlled by lengths L_{IL} and L_{FD}, respectively. Hence, it is inferred that, if these resonance phenomena are appropriately used, the VSWR bandwidth for the IFA with the parasitic IL element can be made wider than that for the IFA without the parasitic element (isolated IFA). This is confirmed in Fig. 4.4, the VSWR bandwidth is more than two times as wide as that of the isolated IFA (increased from approximately 8 to 19%).

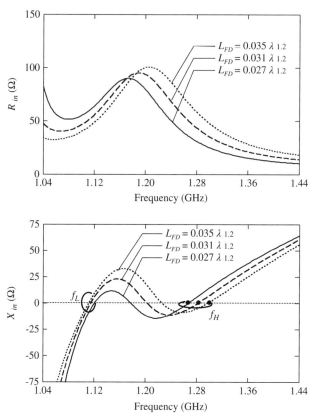

Figurc 4.3   Input impedance $Z_{in} = R_{in} + jX_{in}$ for the IFA with a parasitic IL element, with the IFA feed distance L_{FD} as a parameter. Parameters $(L_{IL}, L_{FD}) = (45.5 \text{ mm, varied}) = (0.182\lambda_{1.2}, \text{varied})$ and those shown in Table 4.1 are used. (Reproduced from Ref. [6] with permission from IET.)

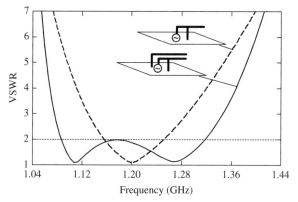

Figure 4.4   VSWR for the IFA with a parasitic IL element, where parameters $(L_{IL}, L_{FD}) = (45.5 \text{ mm}, 7.75 \text{ mm}) = (0.182\lambda_{1.2}, 0.031\lambda_{12})$ and those shown in Table 4.1 are used. (Reproduced from Ref. [6] with permission from IET.)

4.2 INVERTED-F ANTENNA WITH A PAIR OF PARASITIC INVERTED-L ELEMENTS

An IFA with a single parasitic IL element is investigated in Section 4.1. In this section, an IFA with a pair of parasitic IL elements, as shown in Fig. 4.5, is discussed.

Figures 4.6 and 4.7 show the frequency response of the input impedance Z_{in}. In Fig. 4.6, the IL horizontal length L_{IL} is chosen to be a parameter, while the IFA feed distance L_{FD} is fixed: $(L_{IL}, L_{FD}) = (\text{varied}, 8.75\,\text{mm}) = (\text{varied}, 0.035\lambda_{1.2})$. In Fig. 4.7, the IFA feed distance L_{FD} is chosen to be a parameter, while the IL horizontal length L_{IL} is fixed: $(L_{IL}, L_{FD}) = (47.5\,\text{mm}, \text{varied}) = (0.19 \lambda_{1.2}, \text{varied})$. The other parameters used are shown in Table 4.1. The results presented in Figs. 4.6 and 4.7 lead to the conclusion that the lower and higher resonance frequencies are determined by length ABC of the IL and length abc

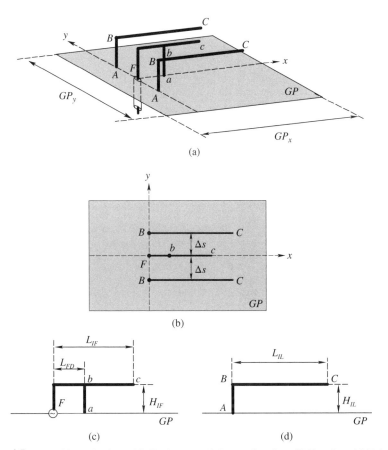

Figure 4.5   IFA with a pair of parasitic IL elements. (a) Perspective view. (b) Top view. (c) Fed inverted F. (d) Parasitic inverted L.

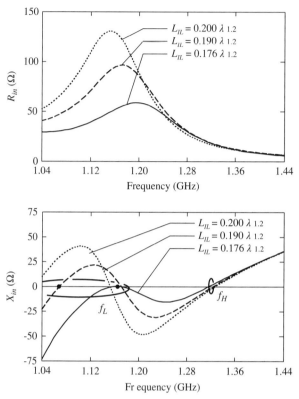

Figure 4.6 Frequency response of the input impedance with the IL horizontal length L_{IL} as a parameter, where the IFA feed distance L_{FD} is fixed: $(L_{IL}, L_{FD}) = (\text{varied}, 8.75 \text{ mm}) = (\text{varied}, 0.035\lambda_{1.2})$. The other parameters used are shown in Table 4.1. (Reproduced from Ref. [6] with permission from IET.)

of the IFA, respectively, as is the case for the IFA with a single parasitic IL element. This facilitates the creation of a wideband VSWR characteristic for the IFA with a pair of ILs. Figure 4.8 shows the frequency response for the VSWR, where parameters $(L_{IL}, L_{FD}) = (47.5 \text{ mm}, 10.75 \text{ mm}) = (0.19\lambda_{1.2}, 0.043\lambda_{1.2})$ are used. The VWSR bandwidth is approximately 26% that is more than three times as wide as that for the isolated IFA.

Figure 4.9 shows the radiation pattern for the IFA with a pair of ILs, where parameters used are the same as those for Fig. 4.8. The radiation is x-polarized, as shown in Fig. 4.9a. The radiation pattern in the y–z plane shown in Fig. 4.9b has two components, E_θ and E_ϕ, resulting from the radiations from the vertical and horizontal wire sections. In the horizontal plane (x–y plane), the radiation component E_θ is almost omnidirectional, as shown in Fig. 4.9c. This is attributed to the fact that the vertical wire sections of the ILs are close (relative to the operating wavelength) to those of the IFA. Finally, Fig. 4.10 shows the gain in the horizontal direction (x-direction) across a frequency range where the VSWR is less than 2. A gain of approximately 5 dBi is obtained.

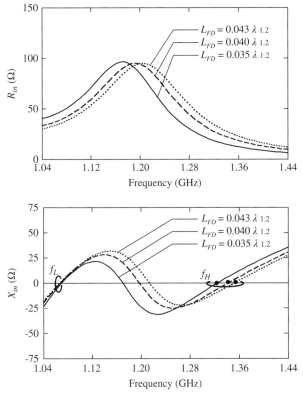

Figure 4.7 Frequency response of the input impedance with the IFA feed distance L_{FD} as a parameter, where the IL horizontal length L_{IL} is fixed: $(L_{IL}, L_{FD}) = (47.5 \text{ mm}, \text{varied}) = (0.19\lambda_{1.2}, \text{varied})$. The other parameters used are shown in Table 4.1. (Reproduced from Ref. [6] with permission from IET.)

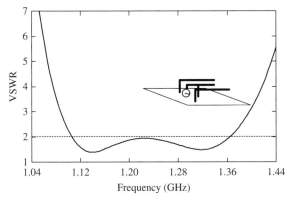

Figure 4.8 Frequency response of the VSWR. Parameters $(L_{IL}, L_{FD}) = (47.5 \text{ mm}, 10.75 \text{ mm}) = (0.19\lambda_{1.2}, 0.043\lambda_{1.2})$ are used, together with the parameters shown in Table 4.1. (Reproduced from Ref. [6] with permission from IET.)

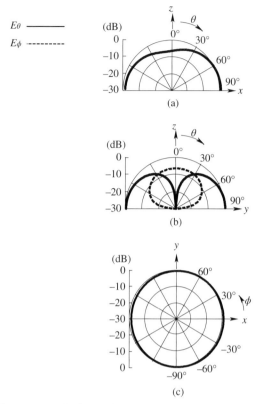

Figure 4.9   Radiation pattern. (a) In the x–z plane. (b) In the y–z plane. (c) In the x–y plane. Parameters $(L_{IL}, L_{FD}) = (47.5\,\text{mm}, 10.75\,\text{mm}) = (0.19\lambda_{1.2}, 0.043\lambda_{1.2})$ are used, together with the parameters shown in Table 4.1. (Reproduced from Ref. [6] with permission from IET.)

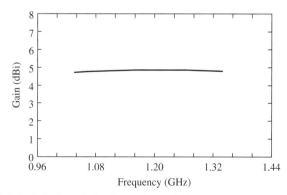

Figure 4.10   Gain in the horizontal direction (x-direction). Parameters $(L_{IL}, L_{FD}) = (47.5\,\text{mm}, 10.75\,\text{mm}) = (0.19\lambda_{1.2}, 0.043\lambda_{1.2})$ are used. The other parameters used are shown in Table 4.1. (Reproduced from Ref. [6] with permission from IET.)

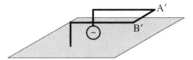

Figure 4.11   Folded inverted-L antenna.

Exercise

Figure 4.11 shows a folded inverted-L (FIL) antenna. Discuss the VSWR bandwidth for this antenna [7,8].

Answer   The equivalent circuit for the input impedance of the FIL is shown in Fig. 4.12 [8], where Z_r is the unbalanced-mode impedance representing the radiation from the FIL, and Z_b is the balanced-mode impedance.

Z_r is expressed by an RLC resonant circuit,

$$Z_r = R + jRQ\left(\frac{f}{f_0} - \frac{f_0}{f}\right)\tag{4.1}$$

where f_0 is the resonance frequency, and R and Q are the radiation resistance and the quality factor, respectively. When the frequency difference $f - f_0 \equiv \Delta f$ is small, Eq. (4.1) can be approximated as

$$Z_r = R + jRQ\left(\frac{2\Delta f}{f_0}\right)\tag{4.2}$$

The balanced-mode impedance Z_b is expressed as

$$Z_b = jZ_c \tan \beta\ell\tag{4.3}$$

where Z_c and β are the characteristic impedance and the phase constant for the balanced-mode line, respectively, and ℓ is the length from the feed point to stub $A'B'$. When ℓ is chosen to be $(2i-1)\lambda/4$ (where $i = 1,2, \ldots$), Eq. (4.3) is approximated as

$$\frac{1}{Z_b} = j\frac{1}{Z_c}\beta_0\ell x\tag{4.4}$$

where $\beta_0 \equiv 2\pi/\lambda_0$ and $x \equiv \Delta f/f_0 \ll 1$ are used. Using Eqs. (4.2) and (4.4), the input admittance $(Y_{in} = 1/Z_{in})$ is expressed as

$$Y_{in} = \frac{1}{2Z_b} + \frac{1}{n^2 Z_r} = \frac{2Z_c R + j\left[n^2 R^2\beta_0\ell x(1 + 4Q^2 x^2) - 4Z_c RQx\right]}{2n^2 Z_c R^2\left(1 + 4Q^2 x^2\right)}\tag{4.5}$$

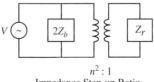

$n^2 : 1$
Impedance Step-up Ratio

Figure 4.12   Equivalent circuit for the folded inverted-L antenna. (Reproduced from Ref. [8] with permission from IEEE.)

The reflection coefficient Γ at the antenna input is expressed using the feed-line admittance for the FIL, Y_0 $(=1/Z_0)$,

$$\Gamma = \frac{Y_0 - Y_{in}}{Y_0 + Y_{in}} \tag{4.6}$$

Using Γ, the voltage standing wave ratio (VSWR), ρ, is expressed as

$$\rho = \frac{1 + |\Gamma|}{1 - |\Gamma|} \tag{4.7}$$

As seen from Eq. (4.5), the susceptance of the admittance Y_{in} becomes zero at three frequencies: let these be $f = f_-, f_0$, and f_+ $(f_- < f_0 < f_+)$. The input impedance $(Z_{in} = 1/Y_{in})$ at each of these frequencies is located on the u-axis of a Smith chart $(u + jv)$. Let Z_{in} be less than Z_0 at $f_0 (Z_{in} < Z_0)$ and greater than Z_0 at f_- and $f_+ (Z_{in} > Z_0)$. In addition, let the VSWRs at these frequencies f_-, f_0, and f_+ be the same: $\rho = \rho_m$. Then,

$$Y_{in} = \rho_m Y_0 \quad \text{at} \quad f = f_0 \tag{4.8}$$

and

$$Y_{in} = \frac{Y_0}{\rho_m} \quad \text{at} \quad f = f_- \text{ and } f = f_+ \tag{4.9}$$

From Eq. (4.5), $Z_{in} = n^2 R$ at $f = f_0$, leading to $\rho_m = 1/(n^2 R Y_0)$.

Y_{in} at $f = f_-$ and f_+ is reduced from Eq. (4.5) to Eq. (4.10)

$$Y_{in} = \frac{1}{n^2 R (1 + Q^2 B^2)} \tag{4.10}$$

Note that $B \equiv 2|x|$, where $|x| = (f_+ - f_0)/f_0 = |f_- - f_0|/f_0$. B is defined as the bandwidth for a VSWR $= \rho_m$ criterion and denoted as BW_{FIL}, where the VSWR is less than or equal to ρ_m across a frequency range of $f_- - f_+$ and greater than ρ_m outside this range.

Substituting Eq. (4.9) into Eq. (4.10), we have

$$\frac{Y_0}{\rho_m} = \frac{1}{n^2 R (1 + Q^2 B^2)} \tag{4.11}$$

Using $\rho_m = 1/(n^2 R Y_0)$, the bandwidth BW_{FIL} is given by

$$BW_{FIL} = 2|x| = \frac{\sqrt{\rho_m{}^2 - 1}}{Q} \tag{4.12}$$

The ratio of BW_{FIL} to the bandwidth for an antenna with a single resonance, BW_{SGL}, is given by

$$BWR \equiv \frac{BW_{FIL}}{BW_{SGL}}$$

$$= \sqrt{\frac{\rho_m(\rho_m + 1)}{\rho_m - 1}} \tag{4.13}$$

where $BW_{SGL} \approx (\rho_m - 1)/Q\sqrt{\rho_m}$ [9] is used. BW_{FIL} for both $\rho_m = 2$ and 3 is approximately 2.5 times as wide as BW_{SGL}. ■

REFERENCES

1. E. Wolff, *Antenna Analysis*, Chap. 3, Nordwood, MA: Artech House, 1988.
2. K. Fujimoto, A. Henderson, K. Hirasawa, and J. James, *Small Antennas*, UK: Research Studies Press, 1987, pp. 116–135.
3. K. Fujimoto and J. James, *Mobile Antenna Systems Handbook*, Nordwood, MA: Artech House, 1994, p. 160.
4. H. Nakano, Y. Wu, H. Mimaki, and J. Yamauchi, An inverted-F antenna with parasitic elements. *Proc. IEICE Soc. Conf.*, September 1994, p. B-35.
5. H. Nakano, R. Suzuki, Y. Wu, H. Mimaki, and J. Yamauchi, Realization of dual-frequency and wide-band VSWR performances using normal-mode helical and inverted-F antennas. *IEEE Trans. Antennas Propag.*, vol. 46, no. 6, pp. 788–793, 1998.
6. H. Nakano, R. Suzuki, and J. Yamauchi, Low-profile inverted-F antenna with parasitic elements on an infinite ground plane. *Proc. IEE Microw. Antennas Propag.*, vol. 145, no. 2, 4, pp. 321–325, 1998.
7. K. Noguchi, Private communication, January 2015.
8. K. Noguchi, M. Mizusawa, M. Nakailama, T. Yamaguchi, Y. Okumura, and S. Betsudan, Increasing the bandwidth of a normal mode helical antenna consisting of two strips. *Proc. IEEE APS Int. Sympo. Antennas Propag.*, vol. 2, pp. 782–785, 1998.
9. D. Sievenpiper, M. Jacob, T. Kanar, S. Kim, J. Long, and R. Quarfoth, Experimental validation of performance limits and design guidelines for small antennas. *IEEE Trans. Antennas Propag.*, vol. 60, no. 1, pp. 8–19, Jan. 2012.

Chapter 5

Multiloop Antennas

A loop antenna composed of a single round or square-shaped wire [1–8] is classi-fied as a resonance antenna. In this chapter, first, we discuss the antenna character-istics of an antenna system composed of N discrete loops [9], each acting as a linearly polarized (LP) element. Second, we discuss a nondiscrete loop antenna system and reveal the radiation characteristics. These discrete and nondiscrete loop antenna systems are used for low-profile base station antennas.

5.1 DISCRETE MULTILOOP (ML) ANTENNAS

A multiloop (ML) antenna is shown in Fig. 5.1, where N discrete loops are located at the same height h. These discrete loops, supported by a spacer of relative permit-tivity ε_r, are backed by a conducting plate. The ith loop ($i = 1, 2, \ldots, N$, counted from the outermost loop) has a square shape of side length S_i (peripheral length of $C_i = 4S_i$). An inverted L-shaped wire, whose vertical and horizontal lengths are, respectively, L_V and L_H, excites the ML without physical contact with the ML con-ductors. This type of excitation was introduced by Nakano, and called *Nakano cou-pling* [10–16]. Note that the bottom end of the L-shaped wire (feed point F) is connected to the inner conductor of a coaxial feed line. The feed point F is on the x-axis: $F(x, y, z) = F((C_N + C_1)/16, 0, 0)$.

The outermost and innermost loop peripheral lengths, C_1 and C_N, are chosen to be slightly larger and smaller than one wavelength, respectively, so that each loop can radiate a broadside beam (the maximum radiation field appears in the z-direction). In this chapter, these peripheral lengths are chosen to be $C_1 = 1.19\lambda_3$ and $C_N = 0.870\lambda_3$ (these are also used later for the modified ML antennas), where λ_3 is the free-space wavelength at a test frequency of 3 GHz. Therefore, the distance between the outermost and innermost loops is $(C_1 - C_N)/8 = 0.04\lambda_3$, and the middle loop peripheral length is $C_{(1+N)/2} = (C_N + C_1)/2 = 1.03\lambda_3$. Note that the relative permittivity ε_r and antenna height h are fixed throughout this chapter: $\varepsilon_r \approx 1$ (honey-comb spacer) and $h = 0.1\lambda_3$.

Low-Profile Natural and Metamaterial Antennas: Analysis Methods and Applications, First Edition.
Hisamatsu Nakano.
© 2016 The Institute of Electrical and Electronics Engineers, Inc. Published 2016 by John Wiley & Sons, Inc.

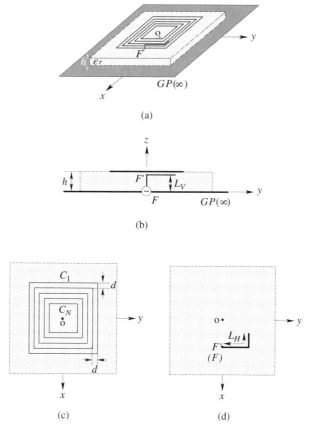

Figure 5.1   Discrete multiloop (ML) antenna. (a) Perspective view. (b) Side view. (c) Top view of discrete loops at $z = h$. (d) Top view of the inverted L-shaped wire at $z = L_v$.

The loop wire radius ρ_{loop} and inverted L wire radius ρ_L are chosen to be the same: $\rho_{loop} = \rho_L = 0.002\lambda_3$ (these values are also used later for the modified ML antennas). The number of loops, N, loop peripheral lengths C_i except for $i = 1$ and N, and center-to-center distance between adjacent wires, d, are varied subject to the objectives of the discussion. C_i and d are a function of N:

$$C_i = [(N - i)C_1 + (i - 1)C_N]/(N - 1) \qquad (5.1)$$

$$d = (C_1 - C_N)/8(N - 1) \qquad (5.2)$$

5.1.1   Antenna Composed of Three Discrete Loops ($N = 3$)

Analysis of the ML antenna is performed by applying the method of moments to an electric field integral equation N2. First, the number of loops is chosen to be $N = 3$. From Eqs. (5.1) and (5.2), the loop circumferences C_i and adjacent wire distance

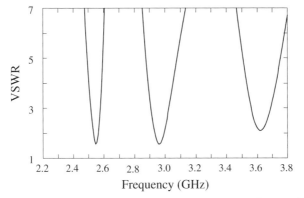

Figure 5.2 VSWR (relative to 50 Ω) for the discrete three-loop antenna as a function of frequency f. (Reproduced from Ref. [9] with permission from IEEE.)

d are given as the following: $(C_1, C_2, C_3, d) = (1.19\lambda_3, 1.03\lambda_3, 0.870\lambda_3, 0.02\lambda_3)$. The length of the inverted L wire is chosen to be $L_V + L_H = 0.051\lambda_3 + 0.199\lambda_3 = \lambda_3/4$.

Figure 5.2 shows the VSWR relative to 50 Ω as a function of frequency f. It is found that a minimum is obtained at each of three frequencies f_i ($i = 1, 2, 3$): $f_1 = 2.552$ GHz, $f_2 = 2.980$ GHz, and $f_3 = 3.624$ GHz. Note that one of the three loops has a loop peripheral length of approximately one wavelength at a given frequency f_i: $C_1 = 1.012\lambda_{2.552}$ at f_1, $C_2 = 1.023\lambda_{2.980}$ at f_2, and $C_3 = 1.051\lambda_{3.624}$ at f_3, where λ_f is the wavelength at frequency f.

The x-directed currents on the loop at each frequency f_i ($i = 1, 2, 3$) are in phase, and hence the maximum radiation field is obtained in the z-direction, with x-directed polarization. Figure 5.3 shows the radiation field, which is decomposed into E_θ and E_ϕ components using the spherical coordinates (r, θ, ϕ). It is found that the radiation patterns for the co-polarization components (E_θ in the $\phi = 0°$ plane and E_ϕ in the $\phi = 90°$ plane) do not significantly change at frequencies f_i ($i = 1, 2, 3$). Further calculation shows that the gain G is approximately 9 dBi at each frequency f_i.

5.1.2 Antennas Composed of Five and Seven Discrete Loops (N = 5 and 7)

Further analysis is performed for discrete five- and seven-loop antennas, for obtaining five and seven minima in the VSWR frequency response, respectively. Again, the same inverted L wire that is used for the discrete three-loop antenna ($L_V + L_H = 0.051\lambda_3 + 0.199\lambda_3 = \lambda_3/4$) feeds these antennas. Figure 5.4a shows the quintuple-frequency operation of the discrete five-loop antenna; the antenna has a VSWR minimum at each of five frequencies f_i ($i = 1, 2, \ldots, 5$). Note that, at a given frequency f_i, one of the five loops has a loop peripheral length of approximately one wavelength. Also, note that the gain at f_i is found to be approximately

$E\theta$ ———
$E\phi$ ---------

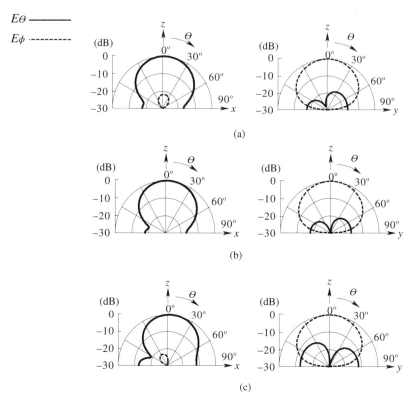

(a)

(b)

(c)

Figure 5.3   Radiation pattern for the discrete three-loop antenna. (a) At $f_1 = 2.552$ GHz. (b) At $f_2 = 2.980$ GHz. (c) At $f_3 = 3.624$ GHz. (Reproduced from Ref. [9] with permission from IEEE.)

$G = 9$ dBi: $(C_1, G) = (1.012\lambda_{2.55}, 8.99$ dBi) at $f_1 = 2.55$ GHz, $(C_2, G) = (1.009\ \lambda_{2.728}, 8.76$ dBi) at $f_2 = 2.728$ GHz, $(C_3, G) = (1.014\lambda_{2.952}, 8.80$ dBi) at $f_3 = 2.952$ GHz, $(C_4, G) = (1.023\lambda_{3.232}, 9.02$ dBi) at $f_4 = 3.232$ GHz, and $(C_5, G) = (1.051\lambda_{3.624}, 8.60$ dBi) at $f_5 = 3.624$ GHz.

The VSWR frequency response for a discrete seven-loop antenna is shown in Fig. 5.4b. It is found that seven minima are realized. The gain for the discrete seven-loop antenna is approximately 9 dBi, as is the case for the discrete three- and five-loop antennas. In other words, the number of loops does not significantly change the gain at the frequency where the VSWR is a minimum.

5.2   MODIFIED MULTILOOP ANTENNAS

Section 5.1 demonstrates that discrete ML antennas have multifrequency operation corresponding to the number of loops N and exhibit an almost constant gain. This section discusses a modified ML antenna, shown in Fig. 5.5. The modified ML

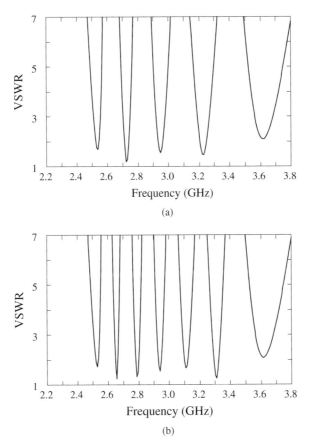

Figure 5.4   VSWR relative to 50 Ω. (a) A discrete multiloop antenna with $N = 5$. (b) A discrete multiloop antenna with $N = 7$. (Reproduced from Ref. [9] with permission from IEEE.)

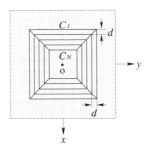

Figure 5.5   Top view of loops with corner-connecting wires.

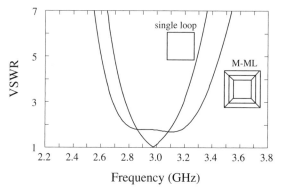

Figure 5.6   VSWR frequency response for a modified three-loop antenna ($N=3$), where the inverted L wire has a total length of more than one-quarter wavelength at 3 GHz ($L_V + L_H = 0.057\lambda_3 + 0.204\lambda_3 = 0.261\lambda_3$). The configuration parameters, including C_1, C_N, ρ_L, h, ε_r, and the coordinates of the feed point F are defined in Section 5.1. The VSWR for a single-loop antenna ($N=1$) is also illustrated. (Reproduced from Ref. [9] with permission from IEEE.)

antenna has the same structure as that shown in Fig. 5.1 except that the corners of the N loops are connected by wires, where the radii of these connecting wires are chosen to be the same as those of the loops (i.e., $\rho_{loop} = 0.002\lambda_3$).

Figure 5.6 shows the VSWR frequency response for $N=3$, where the inverted L wire has a total length of more than one-quarter wavelength at 3 GHz: $L_V + L_H = 0.057 \lambda_3 + 0.204\lambda_3 = 0.261\lambda_3$. For reference, the VSWR for a single loop ($N=1$) is also illustrated. It is found that a smooth frequency response curve is obtained, unlike those shown in Figs. 5.2 and 5.4. This smoothness results from the effect of the connecting

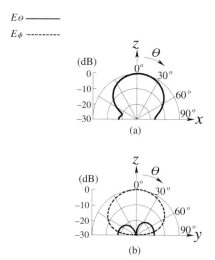

Figure 5.7   Radiation patterns for the modified three-loop antenna ($N=3$) at 3 GHz. (a) In the x–z plane. (b) In the y–z plane. (Reproduced from Ref. [9] with permission from IEEE.)

wires at the four corners of the loops. The bandwidth for a VSWR = 2 criterion of the frequency response in Fig. 5.6 is calculated to be approximately 14.9%.

Further investigation reveals that the VSWR bandwidth is approximately 15.8% for $N=5$ and 15.9% for $N=7$ with the same inverted L wire excitation. These bandwidths are more than 2.5 times as wide as the bandwidth of a single-loop antenna (5.7%).

The gain and radiation pattern for these antennas are also analyzed. The gain for $N=3$, 5, and 7 shows a maximum gain of approximately 9 dBi. The drop from the maximum gain is small (approximately 1 dB) within the frequency bandwidth for a VSWR = 2 criterion. Figure 5.7 shows radiation patterns for $N=3$ at 3 GHz. Note that the radiation patterns for $N=5$ and 7 are similar to those in Fig. 5.7.

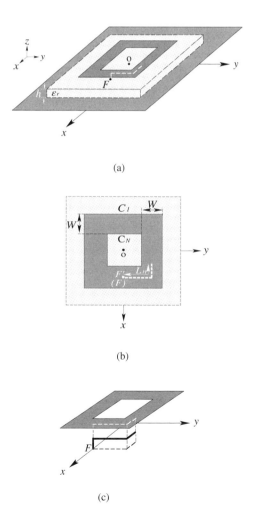

(a)

(b)

(c)

Figure 5.8   Plate-loop antenna. (a) Perspective view. (b) Top view. (c) Bent feed line.

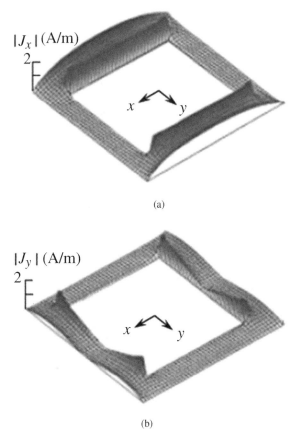

Figure 5.9   Electric current density along the plate loop of the PL antenna at 3 GHz. $\mathbf{J} = J_x\hat{x} + J_y\hat{y}$. (a) | J_x |. (b) | J_y |. (Reproduced from Ref. [9] with permission from IEEE.)

5.3 PLATE-LOOP (PL) ANTENNA

The wide-band VSWR characteristic of a modified ML antenna is discussed in Section 5.2. Figure 5.8 shows a plate-loop (PL) antenna, which corresponds to a structure in the modified ML antenna case where the number of loops, N, is infinite. The PL antenna is excited using an inverted L wire, the same as that used for the modified ML antenna. The analysis of the PL antenna throughout this section is performed using the finite-difference time-domain method described in Section 3.1, where a side length of $\Delta = 0.0065\lambda_3$ is used for the cubic cells within the analysis space.

The width of the PL is chosen to be $W = 0.04\lambda_3$, which is the same as the width used for the ML and modified ML antennas [$W = (C_1 - C_N)/8$, where the peripheral lengths C_1 and C_N are defined in Section 5.1]. Figure 5.9 shows the x- and y-components of the electric current density ($\mathbf{J}_s = J_x\hat{x} + J_y\,\hat{y}$, with $\hat{x}$ and $\hat{y}$ being unit vectors) along the loop at $f = 3$ GHz. It is found that the current is concentrated

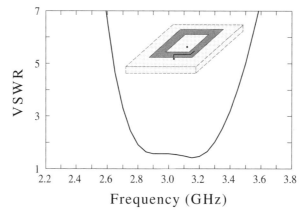

Figure 5.10 VSWR for the PL antenna. (Reproduced from Ref. [9] with permission from IEEE.)

along the conductor edges. Note that this current is obtained using an inverted L feed wire whose length corresponds to approximately one-quarter wavelength at 3 GHz ($L_V + L_H = 0.058\lambda_3 + 0.197\lambda_3 = 0.255\lambda_3$).

It is worthwhile comparing the antenna characteristics of the PL with those of the modified ML. Figure 5.10 shows the VSWR for the PL. The frequency bandwidth for a VSWR = 2 criterion is calculated to be approximately 16.7%, which is slightly larger than the value of 15.9% obtained for a modified ML antenna with $N = 7$. The radiation pattern of the PL is similar to that of the modified ML antenna. Owing to the similarity in the radiation patterns and input impedances of the PL and modified ML antennas, the gain characteristics for these two antennas are also similar. The PL has a gain bandwidth (for a 3-dB gain drop criterion) of approximately 35%, with a maximum gain of approximately 9 dBi.

REFERENCES

1. R. C. Johnson, *Antenna Engineering Handbook*, 3rd edn, Chapter 5, McGraw Hill, NY, 1993.
2. C. Balanis, *Antenna Theory, Analysis and Design*, 3rd edn, Chapter 5, Wiley, 2005.
3. H. Nakano, A numerical approach to line antennas printed on dielectric materials. *Comput. Phys. Commun.*, vol. 68, no 1–3, pp. 441–450, Nov. 1991.
4. P. Deo, A. Mehta, D. Mirshekar-Syahkal, P. Massey, and H. Nakano, Thickness reduction and performance enhancement of steerable square loop antenna using hybrid high impedance surface. *IEEE Trans. Antennas Propag.*, vol. 58, no. 5, pp 1477–1485, 2010.
5. P. Deo, M. Pant, A. Mehta, D. Mirshekar-Syahkal, and H. Nakano, Implementation and simulation of commercial RF switch integration with steerable square loop antenna. *Electron. Lett.*, vol. 47, no. 12, pp. 686–687, June, 2011.
6. K. Hirose, T. Shibasaki, and H. Nakano, Fundamental study on novel loop-line antennas radiating a circularly polarized wave. *IEEE Antennas Wirel. Propag. Lett.*, vol. 11, pp. 476–479, 2012.
7. K. Hirose, T. Shibasaki, and H. Nakano, A loop antenna with parallel wires for circular polarization: its application to two types of microstrip-line antennas. *IEEE Antennas Wirel. Propag. Lett.*, vol. 14, pp. 538–586, 2015.

8. K. Hirose, T. Shibasaki, and H. Nakano, A comb-line antenna modified for wideband circular polarization. *IEEE Antennas Wirel. Propag. Lett.*, vol. 14, pp. 1113–1116, 2015.

9. H. Nakano, M. Fukasawa, and J. Yamauchi, Discrete multiloop, modified multiloop, and plate-loop antennas: multifrequency and wide-band VSWR characteristics. *IEEE Trans. Antennas Propag.*, vol. 50, no. 3, pp. 371–378, 2010.

10. H. Nakano, H. Yoshida, and Y. Wu, C-figured loop antenna. *Electron. Lett.*, vol. 31, no. 9, pp. 693–694, 1995.

11. H. Nakano, M. Yamazaki, and J. Yamauchi, Electromagnetically coupled curl antenna. *Electron. Lett.*, vol. 33, no. 12, pp. 1003–1004, 1997.

12. H. Nakano, S. Tajima, K. Nakayama, and J. Yamauchi, Numerical analysis of honeycomb antennas with an electromagnetic coupling feed system. *IEE Proc. -Microw. Antennas Propag.*, vol. 145, no. 1, pp. 99–103, 1998.

13. K. M. Luk, Y. X. Guo, K. F. Lee, and Y. L. Chow, L-probe proximity fed U-slot patch antenna. *Electron. Lett.*, vol. 34, no. 19, pp. 1806–1807, 1998.

14. C. Mak, K. M. Luk, K. F. Lee, and Y. L. Chow, Experimental study of a microstrip patch antenna with an L-shaped probe. *IEEE Trans. Antennas Propagat.*, vol. 48, no. 5, pp. 777–783, 2000.

15. A. Kishk, R. Chair, and K. F. Lee, Broadband dielectric resonator antennas excited by L-shaped probe. *IEEE Trans. Antennas Propag.*, vol. 54, no. 8, pp. 2182–2189, 2006.

16. Y. Kimura, Y. Shinohe, R. Chayono, and M. Haneishi, Multiband microstrip antennas fed by an L probe. *IEEE AP-S Topical Conf. in Antennas Propag. Wireless Comm. (APWC)*, Torino, September 2011, pp. 796–799.

Chapter 6

Fan-Shaped Antenna

The base station antenna in Chapter 5 forms a unidirectional radiation beam in the direction normal to the antenna plane. The base station antenna in this section, called the fan-shaped antenna [1], is designed to form a conical radiation beam over a frequency range of 3.1–10.6 GHz for use in ultrawideband (UWB) system applications [2–13]. It is found that the antenna height is decreased by increasing the number of fan-shaped elements.

6.1 WIDEBAND INPUT IMPEDANCE

The antenna shown in Fig. 6.1a is composed of two planar conducting arms that are infinitely long and symmetric with respect to the source point. The input impedance of this antenna is frequency-independent [14–16]. An equivalent to this antenna is shown in Fig. 6.1b, where an infinitely long "*single*" arm is used with a ground plane of infinite extent. However, in practice, the single arm cannot be made infinitely long.

Fig. 6.1c shows a triangularly shaped antenna, where the antenna arm and the ground plane shown in Fig. 6.1b have been truncated (the ground plane is circular with diameter D_{GP}). Due to this truncation, the antenna in Fig. 6.1c, called the triangular monopole antenna, no longer has a frequency-independent input impedance characteristic. However, it is expected that the triangular monopole will have a relatively wideband input impedance characteristic, because the antenna structure partially retains the original structure of Fig. 6.1b.

The antenna to be considered in this chapter is shown in Fig. 6.1d; it takes the advantage of the wideband characteristics of the triangular monopole antenna and is designated as the *fan-shaped antenna*. The top circular edge line of the fan-shaped element is defined by radius r_c and inner angle α. The side edge line from point Q $(0, 0, z_Q)$ to point P$(x_p, 0, z_p)$ is defined by an exponential function of $x = -x_0 e^{-t(z-z_p)} + x_0 + x_P$, where $t = (\ln[1+(x_P/x_0)])/(z_P - z_Q)$ and x_0 is a constant that is to be optimized to achieve a low VSWR over the required frequency band.

Low-Profile Natural and Metamaterial Antennas: Analysis Methods and Applications, First Edition. Hisamatsu Nakano.

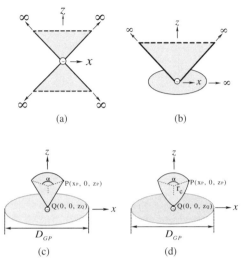

Figure 6.1   Transformation into a fan-shaped structure. (a) Infinitely long two-arm antenna.
(b) Infinitely long single-arm antenna above a ground plane. (c) Triangular monopole antenna with a
round edge. (d) Fan-shaped antenna.

6.2 CHARACTERISTICS OF THE FAN-SHAPED ANTENNA

The structure of the fan-shaped antenna is symmetric with respect to the x–z and y–z planes, where each plane acts as a magnetic wall. The electric field $\mathbf{E}(\mathbf{r}, t)$ and magnetic field $\mathbf{H}(\mathbf{r}, t)$ at symmetric points with respect to the y–z plane are written as

$$E_x(x, y, z, t) = -E_x(-x, y, z, t) \tag{6.1}$$

$$E_y(x, y, z, t) = E_y(-x, y, z, t) \tag{6.2}$$

$$E_z(x, y, z, t) = E_z(-x, y, z, t) \tag{6.3}$$

$$H_x(x, y, z, t) = H_x(-x, y, z, t) \tag{6.4}$$

$$H_y(x, y, z, t) = -H_y(-x, y, z, t) \tag{6.5}$$

$$H_z(x, y, z, t) = -H_z(-x, y, z, t) \tag{6.6}$$

Each of Eqs. (6.1), (6.5), and (6.6) has an odd-symmetry relationship, while each of Eqs. (6.2), (6.3), and (6.4) has an even-symmetry relationship. Similarly, at symmetric points with respect to the x–z plane, each of the E_x, E_z, and H_y components has an even-symmetry relationship, while each of the E_y, H_x, and H_z components has an odd-symmetry relationship.

Analysis is performed using the finite-difference time-domain method (FDTDM) in Section 3.1. Due to the symmetric relationship of the electric and magnetic fields with respect to the x–z and y–z planes, the FDTD calculation can be

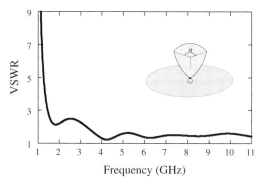

Figure 6.2   Frequency response of the VSWR for a fan-shaped antenna. (Reproduced from Ref. [1] with permission from John Wiley & Sons, Inc.)

performed using just one-fourth of the entire FDTD analysis space of $N_x \Delta x \times N_y \Delta y \times N_z \Delta z$. This leads to a significant reduction in the computation time.

Figure 6.2 depicts the frequency response of the VSWR for a fan-shaped antenna designed using the parameters shown in Table 6.1. It is found that a wideband characteristic for a VSWR $= 2$ criterion is realized across 3.29 GHz, extending to the uppermost analysis frequency (11 GHz). Note that this lower band-edge frequency is further decreased when the number of fan-shaped elements is increased, as shown in the next Section 6.3.

6.3   CROSS FAN-SHAPED ANTENNA (X-Fan Antenna)

The number of fan-shaped elements, N_{FAN}, in Fig. 6.1d is 1. Figure 6.3 shows an extension of the fan-shaped antenna in Fig. 6.1d, where two fan-shaped elements ($N_{FAN} = 2$) intersect at right angles. This antenna is designated as a *cross fan-shaped antenna* (abbreviated as the X-fan antenna). Note that, when N_{FAN} is infinite, the structure constitutes a conducting body of revolution (BOR) antenna, forming a perfectly omnidirectional radiation pattern around the z-axis. The X-fan ($N_{FAN} = 2$) is not a BOR antenna; however, it is expected that the omnidirectionality will be better than that for the original fan-shaped antenna ($N_{FAN} = 1$).

Figure 6.4 shows the radiation pattern of an X-fan, where the parameters used for this calculation are the same as those in Table 6.1. The omnidirectionality and cross-polarization around the z-axis is improved at high frequencies, compared to

Table 6.1   Parameters for a Fan-Shaped Antenna

Symbol	Value	Symbol	Value
λ_4	75 mm	r_c	116Δ
$\Delta(=\Delta_x = \Delta_y = \Delta_z)$	$\lambda_4/400$	$Q(x_Q, y_Q, z_Q)$	$(0,0,1\Delta)$
D_{GP}	320Δ	$P(x_p, y_p, z_p)$	$(100\Delta, 0, 176\Delta)$

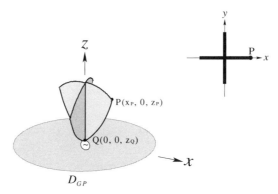

Figure 6.3   Cross fan-shaped antenna (X-fan).

the results for the single fan-shaped antenna shown in Fig. 6.1d. The VSWR for the X-fan is less than 2 across a wide frequency range, including the ultrawideband frequency band, as shown in Fig. 6.5. The antenna height at the lower band-edge frequency (1.85 GHz) is small, approximately $0.27\lambda_{1.85}$, where $\lambda_{1.85}$ is the free-space wavelength at 1.85 GHz.

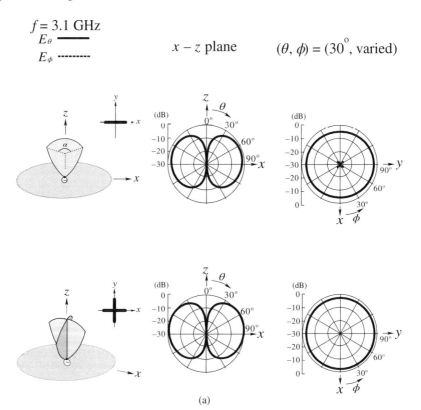

Figure 6.4   Radiation pattern for the X-fan. (a) At 3.1 GHz. (b) At 10.6 GHz. (Reproduced from Ref. [1] with permission from John Wiley & Sons, Inc.)

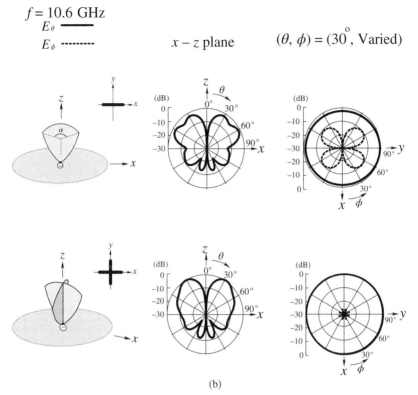

(b)

Figure 6.4   (Continued).

6.4 CROSS FAN-SHAPED ANTENNA SURROUNDED BY A WIRE (X-Fan-W)

Figure 6.6 shows an X-fan antenna surrounded by a conducting wire, which is electrically connected (soldered) to an edge point on each of the four element sections. This antenna is labeled as the X-fan-w. The conducting wire at height $z = z_{wire}$,

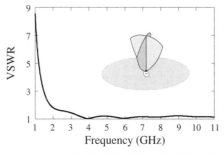

Figure 6.5   VSWR frequency response for the X-fan. (Reproduced from Ref. [1] with permission from John Wiley & Sons, Inc.)

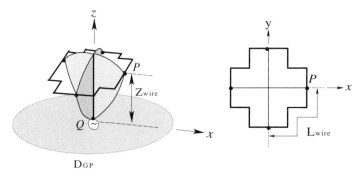

Figure 6.6   X-fan-w, where a conducting wire surrounds the fan-shaped elements.

called the *control wire*, has a total length of $4L_{wire}$. Figure 6.7 shows the VSWR frequency response for the X-fan-w, where the control wire length and height are chosen to be $4L_{wire} = 4 \times 37.5$ mm and $z_{wire} = z_P = 33$ mm, respectively. It is found that the VSWR shows high values around the frequency f_{stop}, that is, a stop band is created. Each current along the control wire segment (of length L_{wire}) exhibits a sinusoidal distribution at f_{stop}; the maximum value of the current appears at the end points of the wire segment and the minimum value (zero) appears at the middle point of the wire segment. It follows that the current along the control wire (of length $4L_{wire}$) has four maxima and four minima (zeroes) at f_{stop}. In other words, a high VSWR is generated when the total length of the control wire at f_{stop} is two guided wavelengths (= approximately two free-space wavelengths). This stop band is used for eliminating interference signals generated by nearby electronic devices.

It is desired that the radiation patterns at frequencies outside the stop band are not affected by the control wire. Figure 6.8 shows the radiation patterns outside the stop band; the X-fan-w retains the inherent radiation pattern of the X-fan. This is attributed to the fact that the main current contributing to the radiation is the current along the edges of the fan-shaped elements (this is also the case for the X-fan antenna).

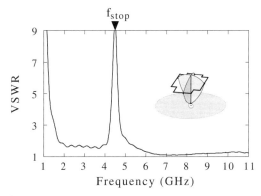

Figure 6.7   VSWR frequency response for the X-fan-w. (Reproduced from Ref. [1] with permission from John Wiley & Sons, Inc.)

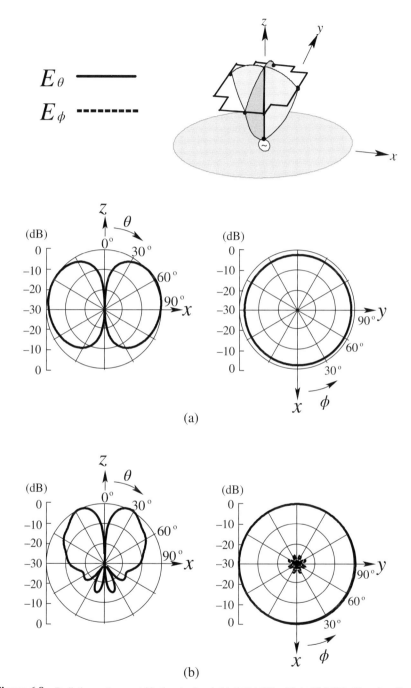

Figure 6.8 Radiation patterns outside the stop band. (a) At 3.1 GHz. (b) At 10.6 GHz. Note that $(\theta, \phi) = (30°,$ varied) in the x-y plane. (Reproduced from Ref. [1] with permission from John Wiley & Sons, Inc.)

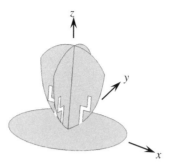

Figure 6.9   X-fan-w, where slots are cut into the fan-shaped elements.

6.5   CROSS FAN-SHAPED ANTENNA WITH SLOTS (X-Fan-S)

A control wire attached to the X-fan generates a stop band in the VSWR frequency response curve, as shown in Fig. 6.7. This section presents a different technique of generating a stop band. Figure 6.9 shows a cross fan-shaped antenna, where slots are cut into the elements. The slots, whose bottom ends are open, have the same length L_{slot}. This antenna is labeled as the X-fan-s to distinguish it from the X-fan-w in Section 6.4.

Figure 6.10 presents the VSWR frequency response for the X-fan-s, with the slot length L_{slot} as a parameter. Other parameters are the same as those for the X fan in Section 6.3. It is found that a stop band is generated when the slot length L_{slot} corresponds to approximately one-quarter of the wavelength. Hence, as L_{slot} is increased, the stop-band center frequency f_{stop} decreases.

The current (absolute value of the current density $\mathbf{J}_s$) on the fan-shaped element at the stop-band center frequency f_{stop} (= 6.88 GHz) is shown in Fig. 6.11a, where the slot length is chosen to be $L_{slot} \approx 11.2$ mm. For comparison, the current

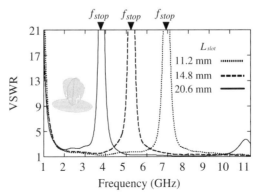

Figure 6.10   VSWR frequency response for the X-fan-s. (Reproduced from Ref. [1] with permission from John Wiley & Sons, Inc.)

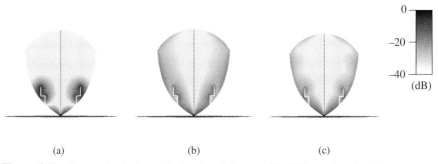

Figure 6.11   Current distribution on the fan-shaped element of the X-fan-s, where L_{slot} is approximately 11. 2 mm. (a) At the stop-band frequency $f_{stop} = 6.88$ GHz. (b) At a low nonresonant frequency of 3.1 GHz. (c) At a high nonresonant frequency of 10.6 GHz.

distributions at two band-edge frequencies of the UWB frequency band, f_L (= 3.1 GHz) and f_H (= 10.6 GHz), both being outside the stop band, are also shown in Fig. 6.11b and c, respectively. It is found that the current is concentrated around the slot at f_{stop} (in other words, the current resonates at f_{stop}), while the currents at frequencies f_L and f_H (nonresonant frequencies) flow along the edges of the fan-shaped element. Consequently, the radiation patterns for the X-fan-s at the frequencies outside the stop band are almost the same as those for the X-fan.

The X-fan-s radiates an omnidirectional or semiomnidirectional beam in the azimuth ϕ plane with an angle θ of between 30° and 60°. The gain in a direction of ($\theta = 60°$, $\phi = 0°$) varies between −0.7 and + 3.2 dBi over a frequency range of 3.1–10.6 GHz, excluding the stop band. It is worth noting that the gain frequency response at nonresonant frequencies for the X-fan-s is similar to those for the X-fan and X-fan-w, resulting from the fact that these three antennas have similar current distributions.

REFERENCES

1. H. Nakano, K. Morishita, Y. Iitsuka, H. Mimaki, T. Yoshida, and J. Yamauchi, Fan-shaped antennas: realization of wideband characteristics and generation of stop bands. *Radio Sci.*, vol. 43, RS4S14, 2008.
2. M. Nakagawa, H. Zhang, and H. Sato, Ubiquitous homelinks based on IEEE 1394 and ultra wideband solutions. *IEEE Commun. Mag.*, vol. 41, pp. 74–82, 2003.
3. H.-J. Park, S.-H. Park, W.-D. Cho, and H.-K. Song, Ubiquitous home networking based on ultra wide bandwidth communication systems. *Second IEEE Workshop on Software Technologies for Future Embedded and Ubiquitous Systems*, May 2004, pp. 127–131.
4. M. Ammann, Square planar monopole antenna. *IEEE National Conference on Antennas and Propagation*, March 1999, pp. 37–40.
5. G. Kumar and K. P. Ray, *Broadband Microstrip Antennas*, Boston: Artech House, 2003.
6. W. L. Stutzman and G. A. Thiele, *Antenna Theory and Design*, 2nd edn, New York: Wiley, 1998.
7. H. G. Schantz and L. Fullerton, The diamond dipole: a Gaussian impulse antenna. *Proc. IEEE AP-S Int. Sympo.*, vol. 4, pp. 100–103, 2001.
8. S. Honda, M. Ito, H. Seki, and Y. Jinbo, A disc monopole antenna with 1: 8 impedance bandwidth and omni-directional radiation pattern. *Proc. ISAP*, September 1992, pp. 1145–1148.

9. H.G. Schantz, Planar elliptical element ultra-wideband dipole antennas. *Proc. IEEE AP-S Int. Sympo.*, vol. 3, pp. 16–21, 2002.

10. T. Yang, S.-Y. Suh, R. Nealy, W. A. Davis, and W. L. Stutzman, Compact antennas for UWB applications. *IEEE Aerosp. Electronic Syst. Mag.*, vol. 19, pp. 16–20, 2004.

11. S.-Y. Suh, W. L. Stutzman, and W. A. Davis, A new ultrawideband printed monopole antenna: the planar inverted cone antenna (PICA). *IEEE Trans. Antennas Propag.*, vol. 52, pp. 1361–1364, 2004.

12. Z. N. Chen, Novel bi-arm rolled monopole for UWB applications. *IEEE Trans. Antennas Propag.*, vol. 53, pp. 672–677, 2005.

13. T. Warnagiris, A monopole with a twist revisited. *Microw. J.*, pp. 54–74, 2005.

14. Y. Mushiake, *Self-Complementary Antennas*, London, UK: Springer, 1996.

15. V. H. Rumsey, *Frequency Independent Antennas*, Academic Press, 1966.

16. H. Nakano, Frequency independent antennas, Chap. 6. in *Modern Antenna Handbook* (ed. C. A. Balanis.), NJ: John Wiley & Sons, 2008.

Chapter 7

BOR–SPR Antenna

$\mathbf{A}$n antenna composed of a conducting body of revolution (BOR) and a conducting parasitic ring shorted to the ground plane (SPR), called the BOR–SPR antenna, is a wideband antenna that generates omnidirectional radiation around the BOR [1–3]. The BOR–SPR has a low-profile structure; the antenna height is approximately 0.07 wavelength at the lower band-edge frequency (f_L) for a VSWR = 2 criterion. (Note: the antenna height is smaller than those for other UWB antennas [4–16].) The diameter of the SPR is small: approximately 0.29 wavelength at f_L. The BOR–SPR has been used as a base station antenna.

7.1 CONFIGURATION

Figure 7.1d shows the steps involved in reaching the final BOR–SPR antenna configuration. The circular patch, shown in Fig. 7.1a, is the radiation element in the first step, where the diameter of the patch (located at height H above the ground plane of diameter D_{GP}) is D_{patch}. A coaxial feed line excites this patch through a vertical narrow conducting strip line of width W_{cent}. The patch in this first step is called the *initial* patch.

A patch with a ring slot, illustrated in Fig. 7.1b, is the radiation element in the second step, where the initial patch of diameter D_{patch} is divided into two sections by a ring slot: the central small patch section called the *patch island* (whose diameter is $2x_1$) and the remaining ring section called the *parasitic ring* (whose inner diameter is $D_{in,\ ring}$). The antenna in the second step is called the patch-slot antenna.

The antenna structure in the third step, illustrated in Fig. 7.1c, differs from the patch-slot antenna structure in that the parasitic ring is short-circuited to the ground plane through four conducting pins, each having width W_{pin}. This antenna is called the patch-slot-pins (PSP) antenna [3]. In the final step, the central section (the patch island and the feed strip of width W_{cent}) for the PSP antenna is replaced with a conducting body of revolution, as illustrated in Fig. 7.1d. The generating line of the BOR is defined by an exponential function, which is given below. This antenna, consisting of the BOR and the shorted parasitic ring (SPR), is called the BOR–SPR antenna.

Low-Profile Natural and Metamaterial Antennas: Analysis Methods and Applications, First Edition.
Hisamatsu Nakano.
© 2016 The Institute of Electrical and Electronics Engineers, Inc. Published 2016 by John Wiley & Sons, Inc.

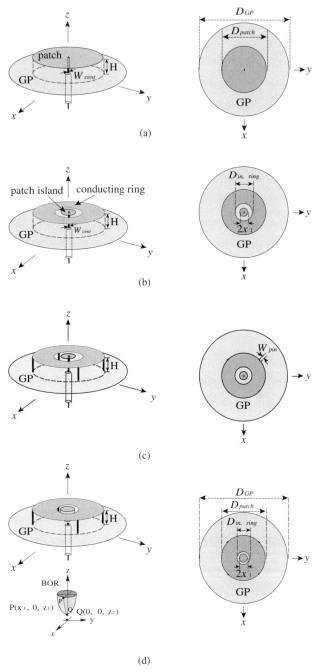

Figure 7.1 Steps involved in reaching the BOR–SPR antenna. (a) Initial patch in the first step. (b) Patch-slot antenna in the second step. (c) Patch-slot-pins antenna in the third step. (d) BOR–SPR antenna.

Each of the antenna structures illustrated in Fig. 7.1 is symmetric with respect to the x–z plane and the y–z plane. The relationship of the electric and magnetic fields, **E** and **H**, at symmetric points with respect to the x–z plane is written as

$$E_x(x, y, z) = E_x(x, -y, z) \tag{7.1}$$

$$E_y(x, y, z) = -E_y(x, -y, z) \tag{7.2}$$

$$E_z(x, y, z) = E_z(x, -y, z) \tag{7.3}$$

$$H_x(x, y, z) = -H_x(x, -y, z) \tag{7.4}$$

$$H_y(x, y, z) = H_y(x, -y, z) \tag{7.5}$$

$$H_z(x, y, z) = -H_z(x, -y, z) \tag{7.6}$$

The relationship at symmetric points with respect to the y–z plane is written as

$$E_x(x, y, z) = -E_x(-x, y, z) \tag{7.7}$$

$$E_y(x, y, z) = E_y(-x, y, z) \tag{7.8}$$

$$E_z(x, y, z) = E_z(-x, y, z) \tag{7.9}$$

$$H_x(x, y, z) = H_x(-x, y, z) \tag{7.10}$$

$$H_y(x, y, z) = -H_y(-x, y, z) \tag{7.11}$$

$$H_z(x, y, z) = -H_z(-x, y, z) \tag{7.12}$$

The antenna is analyzed using the finite-difference time-domain method (FDTDM) described in Section 3.1. The electric and magnetic fields within the entire antenna analysis space are obtained from those within a subspace of ($x \geq 0$, $y \geq 0$) using the relationship in Eqs. (7.1)–(7.12).

7.2 ANTENNA INPUT CHARACTERISTICS OF INITIAL PATCH, PATCH-SLOT, AND PSP ANTENNAS

Figure 7.2a shows the input impedance ($Z_{in} = R_{in} + jX_{in}$) of the initial patch illustrated in Fig. 7.1a, whose configuration parameters are listed in Table 7.1, where the radius of the patch $D_{patch}/2$ corresponds to one-quarter wavelength at frequency 3.75 GHz and the antenna height H corresponds to 0.125 wavelength at the same frequency. It is found that the input reactance at frequencies around 3.75 GHz is highly inductive, resulting in an impedance mismatch to a 50-Ω feed line. Hence, adding capacitance to the antenna is required to reduce this inductive value.

The capacitance is added by cutting a ring slot into the initial patch; see the patch-slot antenna in Fig. 7.1b, where the slot width is $W_{slot} = (D_{in,\ ring} - 2x_1)/2$. The frequency response of the input impedance for the patch-slot antenna is shown

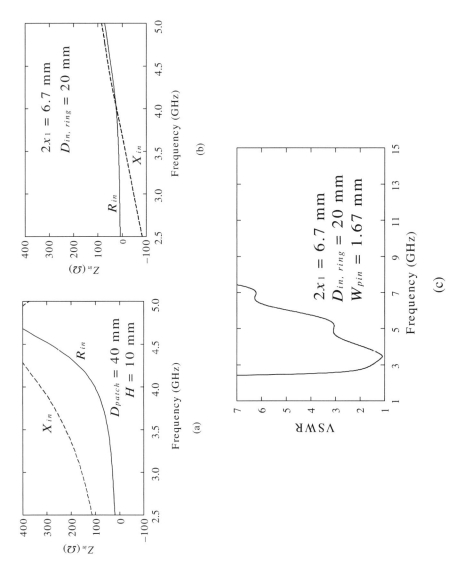

Figure 7.2 Antenna input characteristics. (a) Input impedance for the initial patch antenna. (b) Input impedance for the patch-slot antenna. (c) VSWR for the patch-slot-pins (PSP) antenna. (Reproduced from Ref. [1] with permission from IEEE.)

Table 7.1 Parameters for the Initial Patch Antenna

Symbol	Value	Symbol	Value
D_{GP}	136.7 mm	H	10 mm
D_{patch}	40 mm	W_{cent}	1.7 mm

in Fig. 7.2b. It is confirmed that adding the ring slot contributes to reducing the input reactance X_{in}. However, an issue still exists in that the input resistance R_{in} is less than that required for 50-Ω impedance matching.

This issue is solved by making the conducting part of the patch-slot antenna larger. For this, the patch-slot antenna is transformed into a patch-slot-pins (PSP) antenna, as illustrated in Fig. 7.1c, where conducting pins, each having width W_{pin}, are attached to the periphery of the patch-slot antenna. The antenna input characteristic (VSWR) of this PSP antenna is shown in Fig. 7.2c. It is revealed that the PSP antenna has a wideband VSWR characteristic; approximately 28% bandwidth for a VSWR $= 2$ criterion.

7.3 REPLACEMENT OF THE PATCH ISLAND WITH A CONDUCTING BODY OF REVOLUTION (BOR)

The central section of the PSP antenna illustrated in Fig. 7.1c is composed of a patch island of diameter $2x_1$ connected to a vertical feed strip of width W_{cent}. In the final step of the antenna design, this central section is replaced by a conducting body of revolution (BOR), as illustrated in Fig. 7.1d, in order to obtain a wider VSWR frequency bandwidth. The generating line from the top-edge point $P(x_1, 0, z_1)$ to the bottom point $Q(0, 0, z_2)$ of the BOR is defined by the exponential function used in Section 6.1.

Figure 7.3 shows the frequency response of the VSWR for the BOR–SPR antenna illustrated in Fig. 7.1d, whose parameters are listed in Table 7.2. A VSWR bandwidth of approximately 147% is obtained. The antenna height at the lower

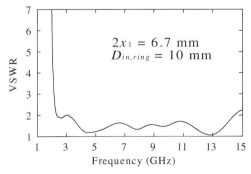

Figure 7.3 Frequency response of the VSWR for the BOR–SPR antenna specified by the parameters shown in Table 7.2. (Reproduced from Ref. [1] with permission from IEEE.)

Table 7.2   BOR–SPR Antenna Parameters

Symbol	Value	Symbol	Value	Symbol	Value
D_{GP}	136.7 mm	H	10 mm	z_1	10 mm
D_{patch}	40 mm	$2x_1$	6.7 mm	z_2	0.8 mm
$D_{in,ring}$	10 mm	W_{pin}	1.67 mm	x_0	1 mm

band-edge frequency is 0.071 wavelength. Detailed calculations reveal that the lower band-edge frequency of the VSWR frequency band for a VSWR = 2 criterion can be shifted downward by appropriately choosing $D_{in, \ ring}$ (i.e., the spacing between the shorted parasitic ring and the BOR). Also, note that the upper band-edge frequency of the VSWR frequency band is not sensitive to the value of the inner ring diameter $D_{in, \ ring}$.

Figure 7.4 illustrates the radiation pattern of the BOR–SPR antenna that has the VSWR frequency response shown in Fig. 7.3. It is found that the BOR–SPR behaves like a monopole above a finite-sized ground plane. The beam direction (BD) is off the z-axis and varies between approximately $\theta = 30°$ and $60°$, where θ is the angle measured from the z-axis. It is also found that, as the frequency decreases, the level of the cross-polarization component in the azimuth plane (fixed at $\theta = 60°$) also decreases.

The absolute gain for the BOR–SPR is shown in Fig. 7.5, where the gain is observed at a fixed direction of $\theta = 60°$. The gain near the lower band-edge frequency of the VSWR frequency band is approximately 2 dBi and the gain near the upper band-edge frequency is approximately 8 dBi. The smallest gain at $f = 8.5$ GHz results from the fact that the radiation in the $\theta = 60°$ direction exhibits a minimal value (valley) in the ripple of the radiation pattern.

Exercise

Figure 7.6 illustrates a BOR–SPR antenna, where the shorted parasitic ring (SPR) is printed on a thin dielectric substrate of relative permittivity ε_r and thickness B. The SPR has four pairs of slots, each having length l_{slot} and width W_{slot} [2]. Investigate the frequency response of the VSWR of this antenna when l_{slot} is changed (use the parameters listed in Tables 7.2 and 7.3). In addition, investigate whether the radiation pattern is influenced by the presence of the slots.

Answer   Figure 7.7 shows the VSWR frequency response, where three slot lengths are used. The VSWR shows a high value at frequencies around a specific frequency (f_{stop}), creating a stop band in the frequency response [2]. This occurs when the slot length corresponds to approximately one-quarter guided wavelength ($\lambda_g/4$). As the slot length is increased, the frequency f_{stop} shifts downward (decreases).

The radiation pattern outside the stop band is not remarkably influenced by the slots; the radiation pattern in the presence of the slots is similar to that in the absence of the slots at frequencies outside the stop band [2]. The stop band can be used to reject interference signals from nearby devices.   ∎

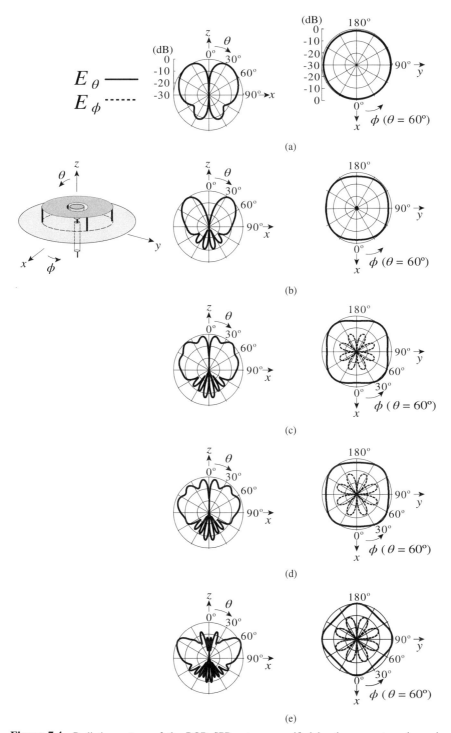

Figure 7.4 Radiation pattern of the BOR–SPR antenna specified by the parameters shown in Table 7.2. The radiation field is normalized to the maximum value in each plane. (a) At 3 GHz. (a) At 6 GHz. (a) At 8.5 GHz. (a) At 9 GHz. (e) At 12 GHz.

101

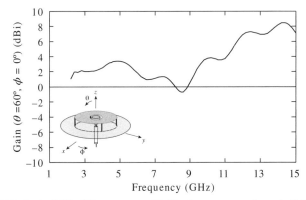

Figure 7.5 Gain for the BOR–SPR antenna specified by the parameters shown in Table 7.2. (Reproduced from Ref. [1] with permission from IEEE.)

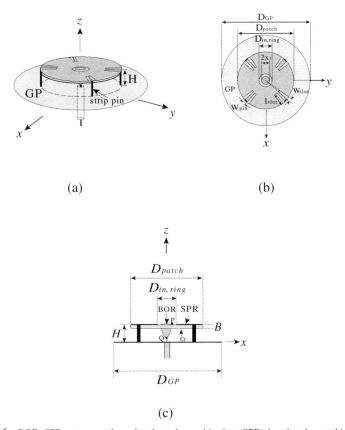

Figure 7.6 BOR–SPR antenna, where the shorted parasitic ring (SPR) is printed on a thin dielectric substrate. Four pairs of slots are cut into the SPR. Each pair of slots is located symmetrically with respect to the pin shorted to the ground plane. (a) Perspective view. (b) Top view. (c) Side view.

Table 7.3   Antenna Parameters

Symbol	Value	Symbol	Value
B	0.83 mm		
ε_r	3.5	W_{slot}	0.83 mm
x_0	0.1 mm		

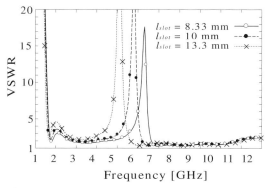

Figure 7.7   VSWR frequency response for $l_{slot} = 8.33$ mm, 10 mm, and 13.3 mm. (Reproduced from Ref. [2] with permission from IEEE.)

REFERENCES

1. H. Nakano, H. Iwaoka, K. Morishita, and J. Yamauchi, A wideband low-profile antenna composed of a conducting body of revolution and a shorted parasitic ring. *IEEE Trans. Antennas Propag.*, vol. 56, no. 4, pp. 1187–1192, 2008.
2. H. Nakano, M. Takeuchi, K. Takeuchi, and J. Yamauchi, Extremely low-profile BOR-SPR-SLOT antenna with stop bands. *IEEE Trans. Antennas Propag.*, vol. 62, no. 6, pp. 2883–2890, 2014.
3. H. Nakano, H. Iwaoka, H. Mimaki, and J. Yamauchi, A wideband PSP antenna radiating a linearly polarized conical beam. In *Proceedings of the 18th International Conference on Applied Electromagnetics and Communications*, Dubrovnik, Croatia, October 2005, pp. 541–544.
4. M. Nakagawa, H. Zhang, and H. Sato, Ubiquitous homelinks based on IEEE 1394 and ultra wideband solutions. *IEEE Commun. Mag.*, vol. 41, pp. 74–82, 2003.
5. H.-J. Park, S.-H. Park, W.-D. Cho, and H.-K. Song, Ubiquitous home networking based on ultra wide bandwidth communication systems. In *Proc. 2nd IEEE Workshop on Software Technologies for Future Embedded and Ubiquitous Systems*, 2004, pp. 127–131.
6. M. Ammann, Square planar monopole antenna. *IEE National Conference on Antennas and Propagation*, March 1999, pp. 37–40.
7. G. Kumar and K. P. Ray, *Broadband Microstrip Antennas*, Boston: Artech House, 2003.
8. W. L. Stutzman and G. A. Thiele, *Antenna Theory and Design*, 2nd edn, New York: Wiley, 1998.
9. H.G. Schantz and L. Fullerton, The diamond dipole: a Gaussian impulse antenna. *Proc. IEEE AP-S Int. Sympo.*, vol. 4, pp. 100–103, 2001.
10. S. Honda, M. Ito, H. Seki, and Y. Jinbo, A disc monopole antenna with 1:8 impedance bandwidth and omni-directional radiation pattern. *Proc. ISAP*, September 1992, pp. 1145–1148.
11. H.G. Schantz, Planar elliptical element ultra-wideband dipole antennas. *IEEE AP-S Int. Sympo.*, vol. 3, pp. 16–21, 2002.

12. T. Yang, S.-Y. Suh, R. Nealy, W. A. Davis, and W. L. Stutzman, Compact antennas for UWB applications. *IEEE Aerosp. Electronic Syst. Mag.*, vol. 19, pp. 16–20, 2004.

13. S.-Y. Suh, W. L. Stutzman, and W. A. Davis, A new ultrawideband printed monopole antenna: the planar inverted cone antenna (PICA). *IEEE Trans. Antennas Propag.*, vol. 52, pp. 1361–1364, 2004.

14. Z. N. Chen, Novel bi-arm rolled monopole for UWB applications. *IEEE Trans. Antennas Propag.*, vol. 53, pp. 672–677, 2005.

15. T. Warnagiris, A monopole with a twist revisited. *Microw. J.*, pp. 54–74, 2005.

16. K.-L. Lau, P. Li, and K.-M. Luk, A monopole patch antenna with very wide impedance bandwidth. *IEEE Trans. Antennas Propag.*, vol. 53, no. 2, pp. 655–661, 2005.

Part II-2

Card Antennas for Mobile Equipment

Chapter 8

Inverted LFL Antenna for Dual-Band Operation

A dual-band antenna called the inverted LFL antenna (InvLFL) is made of a thin conducting film, as shown in Fig. 8.1e [1], where both the top radiation element and the ground plane lie in the same plane, that is, the InvLFL has a card antenna structure. Note that the card structure differs from the layered antenna structure for dual-frequency operation in Ref. 2, where the ground plane backs a radiation element (patch element). It is emphasized that the InvLFL is intended for installation in *mobile* equipment. In such a case, polarization purity (low cross polarization) is not required; however, an appropriate frequency response for the VSWR is important [3–9].

8.1 CONFIGURATION

Figure 8.1 shows the steps involved in reaching the InvLFL configuration shown in Fig. 8.1e. Figure 8.1a is referred to as a modified inverted L element, where the horizontal length and height for the top element are denoted as L'_L and H_L, respectively. Figure 8.1b shows an inverted F radiation element, where height H_F is smaller than height H_L for the inverted L. Part of the horizontal length is denoted as L'_F. Figure 8.1c is a compound radiation element, combining the inverted L in Fig. 8.1a and the inverted F in Fig. 8.1b. Figure 8.1d shows this compound structure with a parasitic inverted L element having length L'_P and height H_P. All structures in Fig. 8.1 are fed from point F, whose distance from the left side edge of the ground plate is L'_{FD}. The ground plate size, $GP_x \times GP_y$, and the strip line width of the radiation elements, w, are fixed (see Table 8.1) for the following discussion.

8.2 DESIGN

The InvLFL is designed for dual-band operation at 2.45 and 5.2 GHz (frequencies used for wireless LAN communications). The analysis for the design is performed

Low-Profile Natural and Metamaterial Antennas: Analysis Methods and Applications, First Edition. Hisamatsu Nakano.

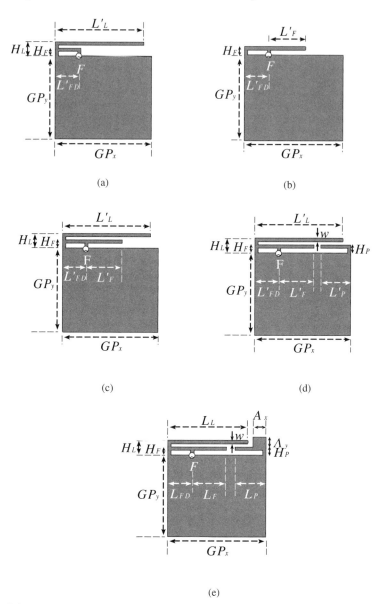

Figure 8.1   Design process. (a) Modified inverted-L element. (b) Inverted-F element. (c) Compound inverted LF. (d) Compound inverted LF with a parasitic inverted-L element. (e) Inverted LFL.

Table 8.1   Parameters

Symbol	Value	Symbol	Value
H_L	4.5 mm	GP_x	30 mm
H_F	2.5 mm	GP_y	25.5 mm
H_P	2.5 mm	w	1 mm

using the finite-difference time-domain method (FDTDM) based on Yee's algorithm described in Section 3.1. The antenna excitation is modeled by a delta-gap voltage source $V_{in}(t)$, which is defined by a sine function modulated by a Gaussian function: $V_{in}(t) = V_{gauss}(t) \sin\omega t$, where $V_{gauss}(t) = \exp[-\{(t-T)/KT\}^2]$, with $K = 0.29$ and $T = 0.646/f_{3dB}$. Note that f_{3dB} is the frequency at which the power spectrum (|Fourier transform of $V_{gauss}(t)|^2$) drops 3 dB from its maximum value.

There are nine structural parameters to be determined for the InvLFL in Fig. 8.1e: the heights (H_L, H_F, H_P), the strip line lengths (L_L, L_F, L_P), the protrusion size (A_x, A_y), and the feed-point location L_{FD}. Let us choose the heights (H_L, H_F, H_P) to be small relative to the wavelengths at 2.45 and 5.2 GHz, as shown in Table 8.1; the antenna height H_L corresponds to approximately 0.037 wavelength at 2.45 GHz, creating a low-profile antenna. The remaining structural parameters (L_L, L_F, L_P), (A_x, A_y) and L_{FD} for the InvLFL are determined through a step-by-step investigation.

Step I   First, the frequency response of the VSWR for the modified inverted L element in Fig. 8.1a is analyzed. To obtain resonance at 2.45 GHz ($\equiv f_{2.45}$), the horizontal length L'_L is chosen such that the total length $H_L + L'_L$ is approximately one-quarter wavelength and then the location of the feed point is adjusted. Figure 8.2 shows the VSWR for $L'_L = 27.5$ mm at frequencies around $f_{2.45}$. Note that the location of the feed point for the remaining steps is fixed at the value of L'_{FD} determined in this first step ($L'_{FD} = 7.5$ mm); however, the horizontal length L'_L is not the final length and is slightly modified in the final step.

Step II Resonance for the inverted F element [Fig. 8.1b] at 5.2 GHz ($\equiv f_{5.2}$) is realized by varying the horizontal length L'_F. When the length $H_F + L'_F$ is approximately one-quarter wavelength at $f_{5.2}$, the VSWR exhibits a desirable characteristic, as shown in Fig. 8.3. Note that the horizontal length $L'_F (= 11$ mm) in this second step is not the final length.

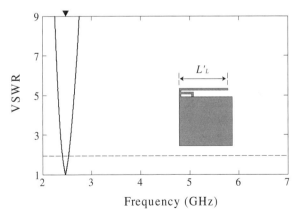

Figure 8.2   VSWR of the modified inverted L at frequencies around 2.45 GHz, where (H_L, L'_L) = (4.5 mm, 27.5 mm) and $L'_{FD} = 7.5$ mm are used. (Reproduced from Ref. [1] with permission from IEEE.)

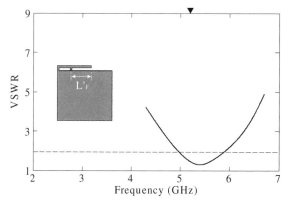

Figure 8.3   VSWR of the inverted F at frequencies around 5.2 GHz, where $(H_F, L'_F) = (2.5\,\text{mm}, 11\,\text{mm})$ and $L'_{FD} = 7.5\,\text{mm}$ are used. (Reproduced from Ref. [1] with permission from IEEE.)

Step III   Investigation is performed to determine whether the compound inverted LF element shown in Fig. 8.1c reproduces the VSWRs shown in Figs. 8.2 and 8.3. Figure 8.4 shows the frequency response of the VSWR for the compound inverted LF. It is found that the VSWR at $f_{2.45}$ remains almost unchanged; however, the VSWR at $f_{5.2}$ deteriorates due to mutual effects between the inverted L and F elements. This deterioration is mitigated in the following step.

Step IV   A parasitic inverted L element is added to the compound inverted LF, as shown in Fig. 8.1d. The height of the parasitic inverted L element is chosen to be equal to the height of the inverted F element: $H_P = H_F$. Figure 8.5 shows the frequency response of the VSWR for three values of the horizontal length of the parasitic inverted-L element, L'_P, where the

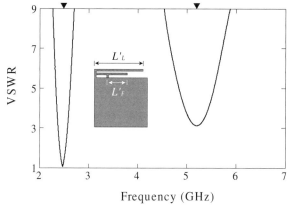

Frequency (GHz)

Figure 8.4   VSWR of the compound inverted LF, where $(H_L, L'_L) = (4.5\,\text{mm}, 27.5\,\text{mm})$, $(H_F, L'_F) = (2.5\,\text{mm}, 11\,\text{mm})$, and $L'_{FD} = 7.5\,\text{mm}$ are used. (Reproduced from Ref. [1] with permission from IEEE.)

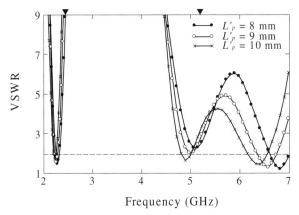

Figure 8.5 VSWR of the compound inverted LF with a parasitic inverted-L element, with the horizontal length L'_P as a parameter. $(H_L, L'_L) = (4.5\,\text{mm}, 27.5\,\text{mm})$, $(H_F, L'_F) = (2.5\,\text{mm}, 11\,\text{mm})$, and $L'_{FD} = 7.5\,\text{mm}$ are used. (Reproduced from Ref. [1] with permission from IEEE.)

structural parameters for the compound inverted LF are held at the same values used in step III. It is revealed that the parasitic inverted L element mitigates the deterioration in the VSWR around 5 GHz observed in step III, generating a resonance between 6 and 7 GHz. Note that the total length, $H_P + L'_P$, is approximately one-quarter wavelength at the frequency where the minimum VSWR for each L'_P appears.

Step V   As shown in Fig. 8.5, the VSWR at frequencies around 5.2 GHz is still larger than 2. This issue is resolved by modifying the antenna structure.

The structure in step IV is modified by (1) reducing the original horizontal strip line lengths L'_L and L'_F and (2) widening the strip width of the parasitic inverted-L element. Note that the widening of the strip width is realized by making a protrusion on the parasitic inverted L element, where part of the strip line is widened to $w + A_y$ over length A_x, as shown in Fig. 8.1e. Figure 8.6 shows the frequency response of the VSWR for the final structure (InvLFL), where lengths L_L and L_F are slightly smaller than the original lengths L'_L and L'_F. This VSWR curve clearly indicates dual-band operation at $f_{2.45}$ and $f_{5.2}$. The frequency bandwidth for a VSWR $= 2$ criterion is approximately 4% for the $f_{2.45}$ band and 32% for the $f_{5.2}$ band. The results are confirmed by experimental results [1].

Figure 8.7 shows the radiation patterns for the InvLFL at frequencies $f_{2.45}$ and $f_{5.2}$. Note that, for confirmation of the FDTDM results, experimental results in the principal planes (x–z and y–z planes) are presented in Ref. 1. Figure 8.8 shows the gain characteristic in the z-direction, G $(\theta = 0°)$; approximately 0.9 dBi at $f_{2.45}$ and 1.7 dBi at $f_{5.2}$, with a small gain variation in each VSWR band. The difference between the gains at $f_{2.45}$ and $f_{5.2}$ (i.e., the gain in the z-direction at 2.45 GHz is smaller than that at 5.2 GHz) is attributed to the following facts: (1) the radiation pattern E_ϕ at 2.45 GHz in each of the x–z and y–z planes is more omnidirectional

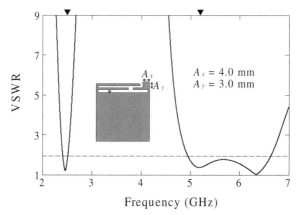

Figure 8.6   VSWR of the InvLFL, where $(H_L, L_L) = (4.5\,\text{mm}, 24.5\,\text{mm})$, $(H_F, L_F) = (2.5\,\text{mm}, 10.5\,\text{mm})$, $L_{FD} = 7.5\,\text{mm}$, $L_P = 9\,\text{mm}$, and $(A_x, A_y) = (4.0\,\text{mm}, 3.0\,\text{mm})$ are used. (Reproduced from Ref. [1] with permission from IEEE.)

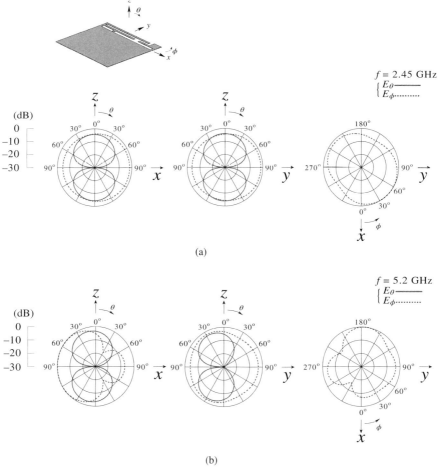

Figure 8.7   Radiation patterns. (a) At $f = 2.45\,\text{GHz}$. (b) At $f = 5.2\,\text{GHz}$. (Reproduced from Ref. [1] with permission from IEEE.)

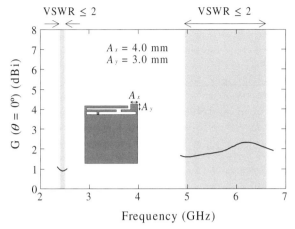

Figure 8.8   Gain in the *z*-direction. (Reproduced from Ref. [1] with permission from IEEE.)

than the pattern at 5.2 GHz; (2) the radiation pattern E_θ at 2.45 GHz in each of the *x*–*z* and *y*–*z* planes (having a figure-eight pattern) shows a wider half-power beam width than the pattern at 5.2 GHz.

Exercise

Design the card antenna shown in Fig. 8.9 such that it operates at four frequencies: 1.7 GHz (LTE band), 2.45 GHz (IEEE 802.11b), 3.5 GHz (4G-advanced LTE), and 5.2 GHz (IEEE 802.11a).

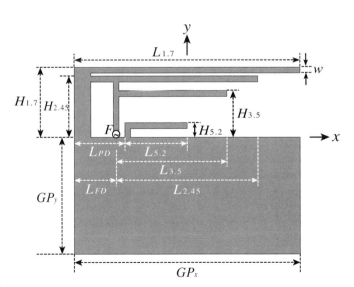

Figure 8.9   Quad-band card antenna.

Table 8.2   Parameters for Fig. 8. 9

Symbol	Value	Symbol	Value	Symbol	Value
$H_{1.7}$	12 mm	$H_{5.2}$	3.5 mm	GP_x	40 mm
$H_{2.45}$	10.5 mm	L_{FD}	7.5 mm	GP_y	20 mm
$H_{3.5}$	8 mm	L_{PD}	9 mm	w	1 mm

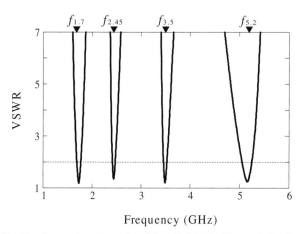

Figure 8.10   Quad-band operation. (Reproduced from Ref. [10] with permission from IEEE), where $L_{FD} = 7.5$ mm $= 0.0425\lambda_{1.7}$, $L_{1.7} = 40$ mm $= 0.227\lambda_{1.7}$, $L_{2.45} = 25$ mm $= 0.204\lambda_{2.45}$, $L_{3.5} = 19.5$ mm $= 0.228\lambda_{3.5}$, and $L_{5.2} = 11$ mm $= 0.191\lambda_{5.2}$.

Use the fixed parameters shown in Table 8.2 [10]. Note that the antenna height $H_{1.7}$ relative to $\lambda_{1.7}$ (free-space wavelength at 1.7 GHz) is chosen to be small: 12 mm $= 0.068\lambda_{1.7}$.

Answer   Heights $H_{1.7}$, $H_{2.45}$, $H_{3.5}$, and $H_{5.2}$ are shown in Table 8.2. Use design steps similar to those discussed in Section 8.2. The result is shown in Fig. 8.10.   ∎

REFERENCES

1. H. Nakano, Y. Sato, H. Mimaki, and J. Yamauchi, An inverted FL antenna for dual-frequency operation. *IEEE Trans. Antennas Propag.*, vol. 53, no. 8, pp. 2417–2421, 2005.
2. K. Wong, *Compact and broadband microstrip antennas*, New York: Wiley, 2002.
3. H. Nakano, T. Kondo, and J. Yamauchi, A card-type, fan-shaped antenna for wideband operation. *Int. J. Microw. Opt Tech.*, vol. 1, no. 1, pp. 100–106, 2006.
4. S. I. Latif, L. Shafai, and S. K. Sharma, Bandwidth enhancement and size reduction of microstrip slot antennas. *IEEE Trans. Antennas Propag.*, vol. 53, no. 3, pp. 994–1003, 2005.
5. J.-Y. Sze and K.-L. Wong, Bandwidth enhancement of a microstrip-line-fed printed wide-slot antenna. *IEEE Trans. Antennas Propag.*, vol. 49, no. 7, pp. 1020–1024, 2001.
6. J.-Y. Jan and J.-W. Su, Bandwidth enhancement of a printed wide-slot antenna with a rotated slot. *IEEE Trans. Antennas Propag.*, vol. 53, no. 6, pp. 2111–2114, 2005.

7. Y.-C. Lin and K.-J. Hung, Compact ultrawideband rectangular aperture antenna and band-notched designs. *IEEE Trans. Antennas Propag.*, vol. 54, no. 11, pp. 3075–3081, 2006.

8. J.-Y. Chiou, J.-Y. Sze, and K.-L. Wong, A broad-band CPW-fed strip-loaded square slot antenna. *IEEE Trans. Antennas Propag.*, vol. 51, no. 4, pp. 719–721, 2003.

9. P. Li, J. Liang, and X. Chen, Study of printed elliptical/circular slot antennas for ultrawideband applications. *IEEE Trans. Antennas Propag.*, vol. 54, no. 6, pp. 1670–1675, 2006.

10. H. Nakano, Y. Kobayashi, and J. Yamauchi, Inverted LFFL card antenna for quadband operation. *Proc. IEEE Int. Sympo. Antennas Propag.*, July 2015, pp. 1712–1713.

Chapter 9

Fan-Shaped Card Antenna

Chapter 6 discusses a self-standing base station antenna consisting of a wideband fan-shaped radiation element above a conducting plate (ground plane). This chapter presents an application of the wideband characteristics of the fan-shaped radiation element to a card antenna, which is abbreviated as the C-FanSA. The C-FanSA is composed of a fan-shaped radiation element and a ground plane, both lying in the same plane, and is designed to operate across a wideband frequency range from 3.1 to 10.6 GHz (UWB). The antenna height of the fan-shaped radiation element above the top edge of the ground plane is chosen to be small (1 cm) for use inside IT mobile devices. The analysis is performed using the finite-difference time-domain method (FDTDM).

9.1 CONFIGURATION

Figure 9.1 shows the configuration of the C-FanSA. The C-FanSA is composed of a fan-shaped element o-P_1-P_2-P_3-o located in the +y space and a ground plane located in the −y space. The fan-shaped element and the ground plane are made of a thin conducting film, and the fan-shaped element itself is sandwiched between thin dielectric layers having relative permittivity ε_r and thickness B. The dielectric is square (area $S \times S$) and the ground plane has area $GP_x \times GP_y$. The C-FanSA is excited by a voltage source between point o (coordinate origin) and point Q, where the distance between o and Q is denoted as Δg.

As shown in Fig. 9.1b, the fan-shaped element is symmetric with respect to the y-axis. The area P_0-P_1-P_2-P_3-P_0 (having an inner angle of 2α) is part of a circle of radius r. The coordinates of points P_m ($m = 0$, 1, 2, and 3) are as follows: $P_0(0, r)$, $P_1(x_1, y_1)$, $P_2(0, 2r)$, and $P_3(−x_1, y_1)$. The edge line from the coordinate origin o to point $P_1(x_1, y_1)$ is defined by the outward exponential function described in Chapter 6.

The configuration parameters are summarized in Table 9.1, where the antenna height (height of the fan-shaped radiation element above the top edge of the ground plane) is chosen to be small: approximately 0.11 wavelength at 3.1 GHz. The

Low-Profile Natural and Metamaterial Antennas: Analysis Methods and Applications, First Edition.
Hisamatsu Nakano.

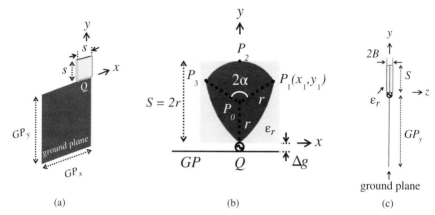

Figure 9.1 Fan-shaped card antenna (C-FanSA). (a) Perspective view. (b) Fan-shaped element. (c) Side view.

remaining parameter, the inner angle 2α, is to be determined such that the C-FanSA realizes a small VSWR over a frequency range of 3.1–10.6 GHz.

9.2 ANTENNA CHARACTERISTICS

Figure 9.2 shows the VSWR relative to 50 Ω, with the inner angle 2α as a parameter. It is found that the C-FanSA with $2\alpha = 120°$ has an extremely wide VSWR response, ranging from 2.9 to more than 10.6 GHz (covering frequencies used in UWB applications). For comparison, the VSWR response for $2\alpha = 120°$ with $\varepsilon_r = 20$ and $\varepsilon_r = 1$ is presented in Fig. 9.3, and shows that the lower band-edge frequency of the VSWR frequency band (for a VSWR = 2 criterion) for $\varepsilon_r = 1$ shifts upward, relative to that for $\varepsilon_r = 20$. This demonstrates the usefulness of sandwiching the fan-shaped radiation element with the dielectric material.

In the following discussion, a fan-shaped element with an inner angle of $2\alpha = 120°$, sandwiched by a dielectric material of $\varepsilon_r = 20$, is used. Figure 9.4 shows the current distribution $J = J_x \hat{x} + J_y \hat{y}$ (where J_x and J_y are the complex number) over the fan-shaped element and ground plane at UWB edge frequencies $f = 3.1$ and 10.6 GHz. It is found that the current over the ground plane is concentrated along the edges. The amplitudes of the current components on the fan-shaped element are

Table 9.1 Antenna Parameters

Symbol	Value	Symbol	Value (mm)
ε_r	20	GP_y	30
$2B$	1 mm	r	5.44
S	10.87 mm	Δg	0.19
GP_x	30 mm		

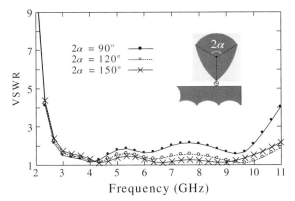

Figure 9.2 Frequency response of the VSWR with the inner angle 2α as a parameter. (Reproduced from Ref. [1] with permission from IJMOT, ISRAMT.)

found to be almost symmetric with respect to the y-axis; $|J_x|$ and $|J_y|$ in the first quadrant $(x > 0,\ y > 0)$ are, respectively, nearly equal to $|J_x|$ and $|J_y|$ in the second quadrant $(x < 0,\ y > 0)$. In addition, detailed analysis reveals that the phase of the J_x components on the fan-shaped element with respect to the y-axis is nearly $180°$ out of phase, while, the phase for J_y components are nearly in-phase.

It is inferred from the aforementioned current distributions that the radiation with polarization that is parallel to the y-axis is expected to be nearly omni-directional around the y-axis. This is confirmed by the radiation pattern shown in Fig. 9.5, where the E_ϕ component, which is parallel to the y-axis in the x–z plane, shows a quasi-omnidirectional pattern (nearly circular pattern). The cross-polarization component E_θ in the x–z plane is due to the current component J_x on the fan-shaped element and the ground plane. Note that, as the frequency increases, the presence of J_x on the ground plane is limited to the area just below the fan-shaped element, and hence E_θ decreases, as can be seen for the radiation pattern in the x–z plane of Fig. 9.5b.

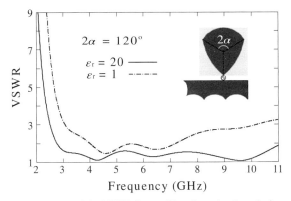

Figure 9.3 Frequency response of the VSWR for $\varepsilon_r = 20$ and $\varepsilon_r = 1$, where the inner angle 2α is $120°$. (Reproduced from Ref. [1] with permission from IJMOT, ISRAMT.)

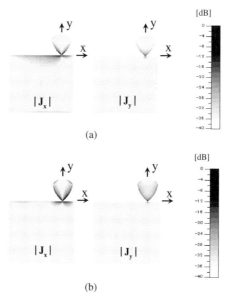

Figure 9.4   Current distribution for $2\alpha = 120°$ and $\varepsilon_r = 20$. (a) At 3.1 GHz. (b) At 10.6 GHz. (Reproduced from Ref. [1] with permission from IJMOT, ISRAMT.)

The frequency response of the gain (not taking into account impedance mismatch) in the z-direction is shown in Fig. 9.6. This has been confirmed by the experimental work in Ref. 1. Figure 9.7 shows an installation example, where two C-FanSAs are installed inside a mobile device.

Exercise

Figure 9.8 shows a two-element antenna array with a decoupling network. Determine two B_1 susceptances and one B_2 shunt susceptance such that port 1 and port 2 are isolated [2].

Answer   The following equations are obtained from Fig. 9.8, where $[Y]$ is the admittance matrix (whose elements are $y_{11} = g_{11} + jb_{11}$, $y_{12} = g_{12} + jb_{12}$, $y_{21} = g_{21} + jb_{21}$, and $y_{22} = g_{22} + jb_{22}$) for the two-element antenna array:

$$\begin{pmatrix} -I_1 \\ -I_2 \end{pmatrix} = [Y]\begin{pmatrix} V_1 \\ V_2 \end{pmatrix} \tag{9.1}$$

$$\begin{pmatrix} I_1 \\ I_2 \end{pmatrix} = [Y_{r11}]\begin{pmatrix} V_1 \\ V_2 \end{pmatrix} + [Y_{r12}]\begin{pmatrix} V_3 \\ V_4 \end{pmatrix} \tag{9.2}$$

$$\begin{pmatrix} I_3 \\ I_4 \end{pmatrix} = [Y_{r21}]\begin{pmatrix} V_1 \\ V_2 \end{pmatrix} + [Y_{r22}]\begin{pmatrix} V_3 \\ V_4 \end{pmatrix} \tag{9.3}$$

$$\begin{pmatrix} I_7 \\ -I_7 \end{pmatrix} = \begin{pmatrix} I_5 - I_3 \\ I_6 - I_4 \end{pmatrix} = [Y_s]\begin{pmatrix} V_3 \\ V_4 \end{pmatrix} \tag{9.4}$$

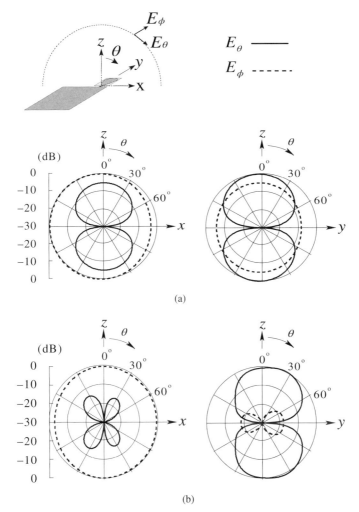

Figure 9.5 Radiation pattern for $2\alpha = 120°$. (a) At 3.1 GHz. (b) At 10. 6 GHz. The configuration parameters are $(\varepsilon_r, 2B, S) = (20, 1 \text{ mm}, 10.87 \text{ mm})$, $(GP_x, GP_y) = (30 \text{ mm}, 30 \text{ mm})$, and $r = 5.44 \text{ mm}$. (Reproduced from Ref. [1] with permission from IJMOT, ISRAMT.)

where

$$[Y_{r11}] = [Y_{r22}] = \begin{pmatrix} jB_1 & 0 \\ 0 & jB_1 \end{pmatrix} \tag{9.5}$$

$$[Y_{r12}] = [Y_{r21}] = \begin{pmatrix} -jB_1 & 0 \\ 0 & -jB_1 \end{pmatrix} \tag{9.6}$$

$$[Y_s] = \begin{pmatrix} jB_2 & -jB_2 \\ -jB_2 & jB_2 \end{pmatrix} \tag{9.7}$$

Using Eqs. (9.1), (9.2), and (9.3), the admittance matrix at plane p_2 is

$$[Y_A] = [Y_{r22}] - [Y_{r21}]([Y] + [Y_{r11}])^{-1}[Y_{r12}] \tag{9.8}$$

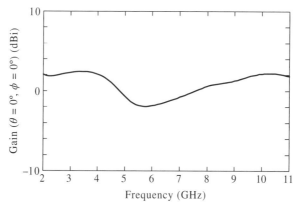

Figure 9.6 Frequency response of the gain in the z-direction. (Reproduced from Ref. [1] with permission from IJMOT, ISRAMT.)

Using Eqs. (9.4) and (9.8), the admittance matrix at plane p_3 is

$$[Y_B] = [Y_A] + [Y_s] \tag{9.9}$$

Port 1 and port 2 are isolated when Y_{B12} is zero. In other words,

$$\text{Re}(Y_{A12}) = 0 \tag{9.10}$$

$$\text{Im}(Y_{A12}) = B_2 \tag{9.11}$$

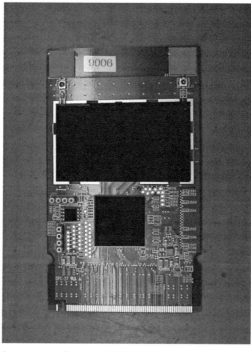

Figure 9.7 Installation example. (Courtesy of Mitsumi Co.)

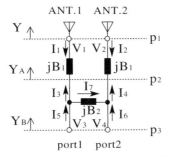

Figure 9.8 Two-element antenna array with a decoupling network. (Reproduced from Ref. [2] with permission from IEEE.)

From Eq. (9.10),

$$B_1 = (-c_1 \pm \sqrt{c_1{}^2 - 4g_{12}c_2})/2g_{12} \qquad (9.12)$$

From Eq. (9.11),

$$B_2 = \frac{B_1^2 g_{12}}{b_{11}g_{22} + g_{11}b_{22} - g_{12}b_{21} - b_{12}g_{21} + B_1(g_{11} + g_{22})} \qquad (9.13)$$

where

$$c_1 = g_{12}(b_{11} + b_{22}) - b_{12}(g_{11} + g_{22}) \qquad (9.14)$$

$$c_2 = -g_{12}(g_{11}g_{22} - b_{11}b_{22} - g_{12}g_{21}) + b_{12}(-b_{11}g_{22} - g_{11}b_{22} + b_{12}g_{21}) \qquad (9.15)$$

∎

REFERENCES

1. H. Nakano, T. Kondo, and J. Yamauchi, A card-type, fan-shaped antenna for wideband operation. *Int. J. Microw. Opt. Tech.*, vol. 1, no. 1, pp. 100–106, 2006.
2. K. Nishimoto, T. Yanagi, T. Fukasawa, H. Miyashita, and K. Konishi, Decoupling networks composed of lumped elements for diversity/MIMO antennas. In: *Proc. ICEAA IEEE-APWC, Torino*, 2013, pp. 307–310.

Chapter 10

Planar Monopole Card Antenna

$\mathbf{T}$he analysis in Ref. 1 revealed that a planar monopole has a wideband characteristic (approximately 50% for a VSWR = 2 criterion). Based on this, a planar monopole card antenna covering a frequency range of 2.4–10.6 GHz (wireless LAN and UWB frequencies) is realized in this chapter [2,3]. The realization is achieved using two planar monopoles: one is a fed monopole and the other is a parasitic monopole. This antenna is abbreviated as the PLMonoP. The design of the PLMonoP is completed in four steps, focusing on the frequency response of the VSWR.

10.1 ANT-1 AND ANT-2

Figure 10.1a shows the antenna structure used in the first step of the design, where a square planar monopole of small area $S \times S$ is located next to the upper edge of a square ground plane ($GP_x = GP_y$). The ground plane and the monopole lie in the same plane, and are backed by a thin dielectric layer of thickness B and relative permittivity ε_r. The monopole is excited from a small gap Δg between the upper edge of the ground plane and the monopole. The antenna in this first step is designated as *Ant*-1; the configuration parameters for *Ant*-1 are summarized in Table 10.1.

Figure 10.1b shows the frequency response of the VSWR for *Ant*-1, where the feed pint F is located at distance L'_{FD}, measured from the left edge of the ground plane. As L'_{FD} is decreased, the lower band-edge frequency of the VSWR (for a VSWR = 2 criterion), f_L, decreases. However, *Ant*-1 does not cover the target frequency range of 2.4–10.6 GHz, and hence the antenna structure must be modified.

Before making the modification, the current distribution on *Ant*-1 is investigated. Figure 10.1c shows the surface current $\mathbf{J}_s$ on the planar monopole and the ground plane. It is found that, as L'_{FD} is decreased, the current along the left edge of the ground plane increases. This situation is undesirable in hand-set devices (when the device is held, the proximity of the hand changes the current distribution and the input impedance).

Low-Profile Natural and Metamaterial Antennas: Analysis Methods and Applications, First Edition. Hisamatsu Nakano.
© 2016 The Institute of Electrical and Electronics Engineers, Inc. Published 2016 by John Wiley & Sons, Inc.

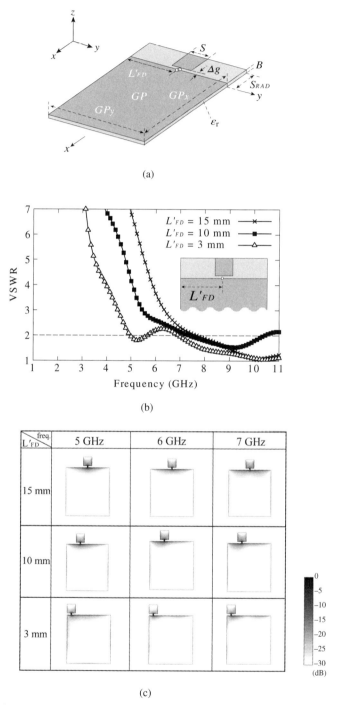

(a)

(b)

(c)

Figure 10.1 Ant-1. (a) Configuration. (b) Frequency response of the VSWR with the feed location L'_{FD} as a parameter. (c) Current distribution. Note that $15\,\text{mm} = GP_x/2$, $10\,\text{mm} = GP_x/3$, and $3\,\text{mm} = GP_x/10$. (Reproduced from Ref. [3] with permission from IEEE.)

Table 10.1 Configuration Parameters for Ant-1

Symbol	Value	Symbol	Value
S	6 mm	GP_x	30 mm
B	1 mm	GP_y	30 mm
ε_r	4.2	Δg	1 mm

Source: Reproduced from Ref. 3 with permission from IEEE.

The second step is to reduce the lower band-edge frequency f_L obtained in the first step. For this, a parasitic planar monopole is added to the fed monopole, as shown in Fig. 10.2a. This parasitic monopole has the same size as the fed monopole and is shorted to the ground plane. This antenna in Fig. 10.2a is designated as *Ant-2*. Based on the results shown in Fig. 10.1c, the feed location is chosen to be $L'_{FD} = GP_x/3$ to avoid an increase in the edge current of the ground plane. Figure 10.2b shows the frequency response of the VSWR for $L'_{FD} = GP_x/3$. As desired, the lower band-edge frequency f_L is reduced to approximately 4 GHz. This reduction in f_L is attributed to the induced current on the parasitic monopole, as shown in Fig. 10.2c, resulting in an increase in the radiation area. However, the target lower band-edge frequency f_L (2.4 GHz) is still not obtained.

10.2 ANT-3 AND ANT-4

In the third step, the lower band-edge frequency f_L in the second step is further reduced by increasing the height of the parasitic monopole. Figure 10.3a shows the antenna structure for the third step, where the height of the parasitic monopole above the upper edge of the ground plane, S_{RAD}, is increased from that used for *Ant-2*. This antenna is designated as *Ant-3*. The VSWR frequency response for *Ant-3* is presented in Fig. 10.3b, together with the frequency response for *Ant-2*. The lower band-edge frequency f_L is found to decrease to approximately 3 GHz. It is emphasized that *Ant-3* realizes a low VSWR over the frequency range used for UWB systems (3.1–10.6 GHz). However, *Ant-3* does not satisfy the design goal yet, that is, it does not have a low VSWR at frequencies around 2.4 GHz (LAN frequencies). This issue is resolved in the fourth step.

Figure 10.4a shows the antenna used in the fourth step; a horizontal strip of length L is added to the parasitic monopole of *Ant-3*. This antenna is designated as *Ant-4*. A parametric study reveals that there exists an optimum length L that allows for operation at LAN frequencies. As seen from Fig. 10.4b, the optimization leads to a VSWR of less than 2 over a frequency range of 2.3–11 GHz. Thus, the design specification for the VSWR is satisfied and *Ant-4* becomes the final design for the PLMonoP card antenna. The top element size of the PLMonoP is small: $S_{RAD} \times (L + S) = 11 \text{ mm} \times 13 \text{ mm} = (0.09 \text{ wavelength}) \times (0.10 \text{ wavelength})$ at 2.4 GHz. The radiation pattern at 2.45 GHz for the PLMonoP card antenna is presented in Fig. 10.4c.

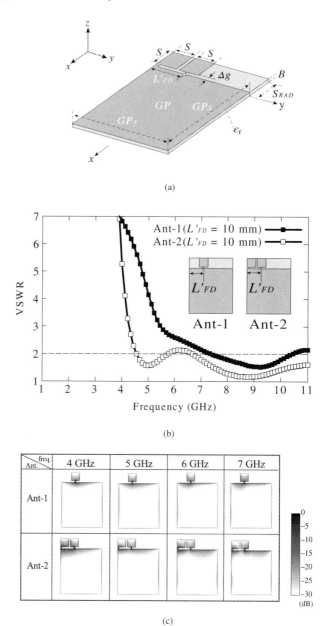

Figure 10.2   Ant-2.(a) Configuration. The feed location is $L'_{FD} = GP_x/3 = 10\,\text{mm}$. (b) Comparison of the VSWR frequency response for *Ant*-2 and *Ant*-1. (c) Comparison of the current distribution for *Ant*-2 and *Ant*-1. (Reproduced from Ref. [3] with permission from IEEE.)

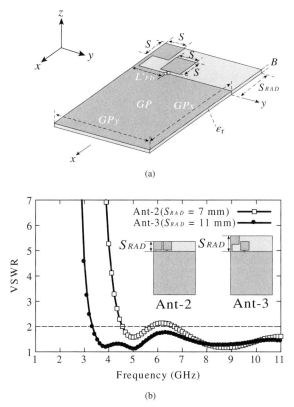

(a)

(b)

Figure 10.3   Ant-3.(a) Configuration, where the feed location is $L'_{FD} = GP_x/3 = 10$ mm. (b) Frequency response of the VSWR, where the height of the parasitic element is $S_{RAD} = 11$ mm. For comparison, the frequency response of the VSWR for Ant-2 is also presented. (Reproduced from Ref. [3] with permission from IEEE.)

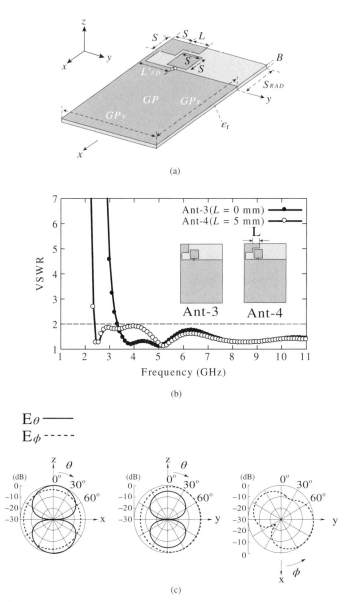

(a)

(b)

(c)

Figure 10.4   Ant-4. (a) Configuration, where $L'_{FD} = GP_x/3 = 10$ mm $\equiv L_{FD}$, $S_{RAD} = 11$ mm, $L = 3$ mm. (b) Frequency response of the VSWR. For comparison, the frequency response of the VSWR for *Ant*-3 is also presented. (c) Radiation pattern for *Ant*-4 at 2.45 GHz. (Reproduced from Ref. [3] with permission from IEEE.)

REFERENCES

1. T. Maruyama, J. Yamauchi, and H. Nakano, Transformation of a plate-type antenna into skeleton-type antennas. *Proc. IEICE Commun. Soc. Conf.*, 2006, p. 1–118,
2. Y. Asano, J. Yamauchi, and H. Nakano, A card-type wide band antenna. *Proc. IEICE Gen. Conf.*, 2005, p. 1–76.
3. H. Nakano, T. Igarashi, Y. Iitsuka, and J. Yamauchi, Single wideband card antenna for LAN and UWB systems, *iWAT*, SS4-3, Lisbon, March 2010, pp. 1–4.

Part II-3

Beam forming Antennas

Chapter 11

Inverted-F Antenna Above an Electromagnetic Band-Gap Reflector

An inverted-F antenna located above a plane reflector (a ground plane) that is assumed to be a perfect electric conductor (PEC) is discussed in Chapter 4, where the antenna height above the ground plane corresponds to approximately 0.09 wavelength. It is noted that, as the antenna height is further decreased, input impedance matching becomes difficult. In this chapter, the PEC reflector is replaced with an electromagnetic band-gap (EBG) reflector [1] to realize a smaller antenna height. The radiation characteristics for realizing a beam-scanning antenna are discussed.

11.1 INVERTED-F ARRAY WITH AN EBG REFLECTOR (EBG–INVF ARRAY)

Figure 11.1a shows an array antenna composed of four identical inverted F (InvF) elements that are specified by strip width W, vertical length L_V, horizontal length L_H, and distance D_{BS} [2,3]. Two InvFs are arrayed on the x-axis, with a spacing of d; the other two InvFs are arrayed on the y-axis, with the same spacing d. Excitation for this array is applied at points (F_{-x}, F_{+x}, F_{-y}, F_{+y}).

An EBG reflector that is square and composed of N^2 patches printed on a dielectric substrate (relative permittivity ε_r and thickness B) is located behind the InvFs. Each patch in the reflector is square with side length S_{patch}; the spacing between neighboring patches is δ. Note that the EBG side length S_{EBG} is given as $S_{EBG} = (N-1)\delta + N S_{patch}$; the distance from the surface of the EBG reflector to the horizontal element of the InvF is $L_v - B \equiv \Delta h$; and the distance from the bottom of the EBG reflector to the horizontal element of the InvF is $L_v + W \equiv H$.

The array in Fig. 11.1 is designated as the EBG–InvF array antenna. Design of the EBG–InvF array is implemented for operation at frequencies around 6 GHz. For this, the following (fixed) parameters are used [4]: $W = W_{fed} = 0.03\lambda_6$, $L_V = B = 0.04$

Low-Profile Natural and Metamaterial Antennas: Analysis Methods and Applications, First Edition.
Hisamatsu Nakano.

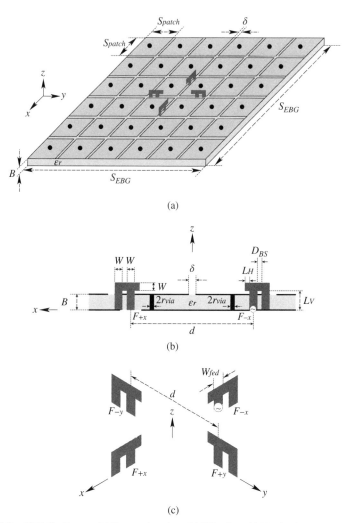

Figure 11.1 EBG–InvF array. (a) Perspective view. (b) Side view. (c) Feed points.

λ_6, $D_{BS} = 0.02\lambda_6$, $d = 0.392\lambda_6$, $\varepsilon_r = 2.2$, $S_{patch} = 0.246\lambda_6$, $2r_{via} = 0.048\lambda_6$, and $\delta = 0.03\lambda_6$, where λ_6 is the free-space wavelength at 6 GHz. The horizontal length L_H and number of patches N^2 (and hence the EBG side length S_{EBG}) are varied subject to the objectives of the discussion.

11.2 ANTENNA CHARACTERISTICS

A situation is considered where point F_{-x} is excited and the other three points (F_{+x}, F_{-y}, F_{+y}) are left open-circuited. The antenna characteristics for this situation are investigated using the finite-difference time-domain method (FDTDM) described in Chapter 3.1 and a commercially available EM solver (CST MW Studio [5]).

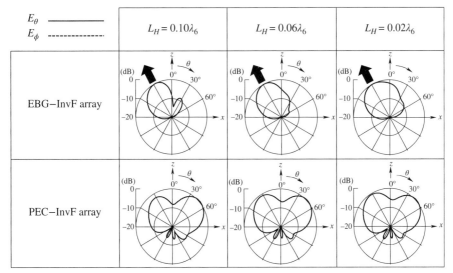

Figure 11.2 Radiation pattern at 6 GHz when the horizontal length L_H is changed. The number of patches is $N^2 = 36$ and the EBG reflector side length is $S_{EBG} = 1.626\lambda_6$.

Figure 11.2 shows behavior of the radiation pattern at 6 GHz when the horizontal length L_H is changed, where the number of patches is held at $N^2 = 36$ (which leads to an EBG reflector side length of $S_{EBG} = 1.626\lambda_6$). It is revealed that the radiation pattern is tilted in the x–z plane. For comparison, the radiation pattern of a corresponding array backed by a PEC reflector (PEC–InvF array) is also shown in this figure, where the PEC side length S_{PEC} is the same as the EBG side length: $S_{PEC} = S_{EBG} = 1.626\lambda_6$. It is found that there is a remarkable difference in the radiation patterns of the EBG–InvF and PEC–InvF arrays.

The tilt in the beam for the EBG–InvF array in the x–z plane is attributed to the x-directed current on the surface of the EBG reflector, J_x. As shown in Fig. 11.3a, the amplitude of the current in the negative x-region (shadowed region) is larger than that in the positive x-region; in addition, as shown in Fig. 11.3b, the phase delay of the current in the negative x-region increases toward the edge of the EBG reflector. Consequently, the maximum radiation appears in the negative x-space.

Figure 11.4 shows the effect of the number of patches N^2 on the radiation pattern at 6 GHz, where the horizontal length is held at $L_H = 0.02\lambda_6$. As the number of patches increases, the radiation beam becomes narrower. The HPBW of the radiation beam in the x–z plane is calculated to be approximately 78° for $N^2 = 16$, 63° for $N^2 = 36$, and 48° for $N^2 = 64$.

The EBG reflector works as a perfect magnetic conductor (PMC) within a very narrow frequency region around 6 GHz. This affects the input impedance (and hence the VSWR). Figure 11.5 shows the VSWR frequency response for a representative case where the structural parameters are $(L_H, N^2) = (0.02\lambda_6, 36)$. A desirable small VSWR is obtained around 6 GHz, with a 7.5% bandwidth for a VSWR = 2 criterion.

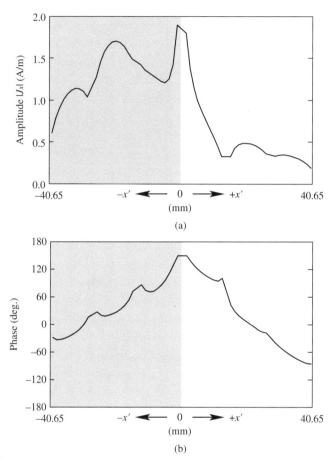

Figure 11.3   The *x*-directed current on the surface of the EBG reflector [along the *x*'-axis (at *z* = B) near the *x*-axis] at 6 GHz, where L_H is 0.02λ_6 and N^2 = 36. (a) Amplitude. (b) Phase.

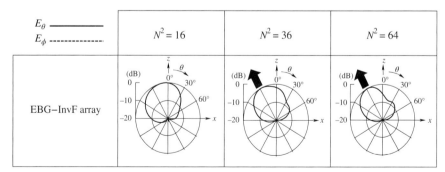

Figure 11.4   Radiation pattern at 6 GHz when the number of patches N^2 is changed, where L_H = 0.02λ_6. (a) N^2 = 16. (b) N^2 = 36. (c) N^2 = 64.

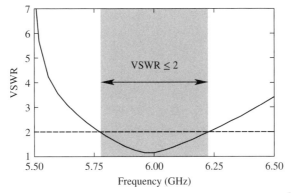

Figure 11.5 VSWR frequency response, where the structural parameters are $(L_H, N^2) = (0.02\lambda_6, 36)$.

The EBG–InvF array is symmetric with respect to the center of the EBG reflector. Therefore, the EBG–InvF array can produce a tilted beam in the positive x-direction by exciting point F_{+x} and leaving the remaining points (F_{-x}, F_{+y}, F_{-y}) open-circuited. Similarly, a tilted radiation beam in the $+y$ or $-y$ direction can be obtained by exciting point F_{+y} or F_{-y}, respectively. These changes in the excitation at points $(F_{+x}, F_{-x}, F_{+y}, F_{-y})$ are performed using switching circuits. Note that the VSWR frequency response obtained for the F_{-x} excitation is reproduced for the other three excitations.

Exercise

Design an EBG–InvF array that radiates reconfigurable six beams.

Answer Figure 11.6 shows an EBG–InvF array, where 37 hexagonal patches are used [6,7]. The radiation pattern and the VSWR (when the feed point for the 1st antenna is

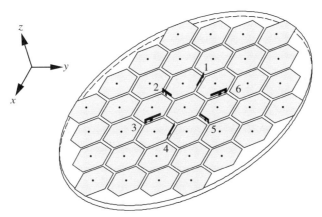

Figure 11.6 Array composed of six InvF elements with an EBG reflector.

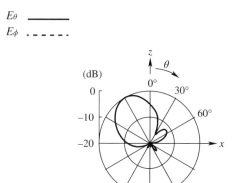

Figure 11.7   Radiation pattern at 6 GHz.

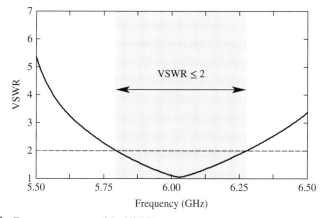

Figure 11.8   Frequency response of the VSWR.

excited and the other feed points are open-circuited) are presented in Figs. 11.7 and 11.8, respectively. The parameter are as follows: $W = 0.0200\lambda_6$, $W_{fed} = 0.0254\lambda_6$, $L_V = 0.0200\lambda_6$, $L_H = 0.0680\lambda_6$, $D_{BS} = 0.0313\lambda_6$, $d = 0.4636\lambda_6$, $\varepsilon_r = 2.2$, $B = 0.0400\lambda_6$, $S_{patch} = 0.1400\lambda_6$, $2r_{via} = 0.0200\lambda_6$, and $\delta = 0.0200\lambda_6$, where λ_6 is the free-space wavelength at 6 GHz. ■

REFERENCES

1. D. Sievenpiper, L. Zhang, R. Broas, N. Alexopoulos, and E. Yablonovitch, High-impedance electromagnetic surfaces with a forbidden frequency band. *IEEE Trans. Microw. Theory Tech.*, vol. 47, no. 11, pp. 2059–2074, 1999.
2. H. Nakano, Y. Asano, and J. Yamauchi, A wire inverted F antenna on a finite-sized EBG material. In: *Proc. IEEE International Workshop on Antenna Technology*, Singapore, March 2005, pp. 13–16.
3. H. Nakano, Y. Asano, H. Mimaki, and J. Yamauchi, Tilted beam formation by an array composed of strip inverted F antennas with a finite-sized EBG reflector. In: *Proc. IEEE Int. Symp. Microw., Antenna, Propag. and EMC Techno., MAPE 2005*, vol. 1, August 2005, pp. 438–441.

4. R. Kato, J. Yamauchi, and H. Nakano, Effects of pins on the antenna characteristics of an array composed of four inverted F elements for a tilted beam. In: *Proc. IEICE Society Conference*, Sendai, September 2015, p. B-1-68.
5. https://www.cst.com/Products/CSTMWS.
6. R. Okamura, J. Yamauchi, and H. Nakano, Six-direction beam-steerable antenna with an EBG reflector. In: *Proc. IEICE General Conference*, Kusatsu, March 2015, p. B-1-41.
7. R. Okamura, J. Yamauchi, and H. Nakano, Effects of an EBG reflector plate on an inverted-F antenna array. In: *Proc. IEICE Society Conference*, Sendai, September 2015, p. B-1-104.

Chapter 12

Reconfigurable Bent Two-Leaf and Four-Leaf Antennas

This chapter presents two low-profile antennas that radiate a *wide* unidirectional beam in the horizontal plane: one antenna is designated as the BeToL [1] and the other is designated as the BeFoL [2]. These antennas are reconfigurable, that is, the direction of the unidirectional beam can be changed using switching circuits [3–5]. In addition, an array antenna composed of BeToL elements is investigated to form a fan beam [9]. The analysis in this chapter is performed using the finite-difference time-domain method (FDTDM).

12.1 BeToL ANTENNA

The omnidirectional radiation from a z-directed dipole can be transformed into unidirectional radiation toward the positive x-space by installing a conducting plane reflector, as shown in Fig. 12.1. In this case, the radiation on the $\pm y$-axes is zero because there is a phase difference of $180°$ between the current along the dipole and its image. Hence, the radiation pattern (of z-directed electric field) in the x–y plane is not wide. This section presents an antenna that resolves this narrow beam issue and realizes a reconfigurable wide beam.

12.1.1 Configuration

Figure 12.2 shows an antenna composed of leaf-like conducting strips A and B above a ground plane (GP) of area $GP_x \times GP_y$. The antenna is designated as the bent two-leaf antenna (BeToL). The BeToL has a volume of $S_x \times S_y \times S_z$ above the ground plane. Point a of leaf A is fed and point b of leaf B is shorted to the ground plane.

Each leaf has a triangular section of height t and bottom $\Delta_{btm} = S_y$. The vertical section of each leaf is composed of a narrow strip (length Δh and width ΔW_v) and a

Low-Profile Natural and Metamaterial Antennas: Analysis Methods and Applications, First Edition.
Hisamatsu Nakano.
© 2016 The Institute of Electrical and Electronics Engineers, Inc. Published 2016 by John Wiley & Sons, Inc.

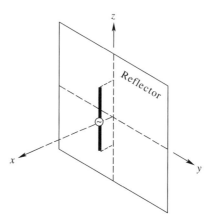

Figure 12.1   Dipole backed by a conducting plane.

wide strip (length $S_z - \Delta h$ and width W_v). The length measured along the centerline of the conducting leaf, $S_z + t$, is approximately one-quarter wavelength. For realizing a low-profile antenna, the antenna height S_z is chosen to be small.

The following parameters are fixed throughout the BeToL design: $S_z = 0.12\lambda_3$, $t = 0.1\lambda_3$, $(\Delta W_v, \Delta h) = (0.02\lambda_3, 0.01\lambda_3)$, where λ_3 is the free-space wavelength at 3 GHz $\equiv f_3$. The major parameters are (W_v, S_x, S_y) and are varied subject to the objectives of the analysis. Note that the ground plane, unless otherwise mentioned, is assumed to be of infinite extent: $GP_x = \infty$ and $GP_y = \infty$.

12.1.2 Radiation Pattern

Fed leaf A excites parasitic leaf B and produces a current on leaf B, which contributes to forming a unidirectional radiation pattern, together with the current on leaf A.

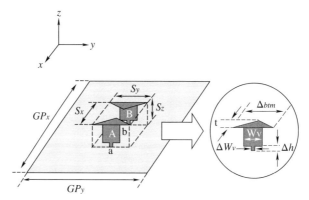

Figure 12.2   Bent two-leaf (BeToL) antenna. Leaf A is fed and leaf B is parasitic and shorted to the ground plane.

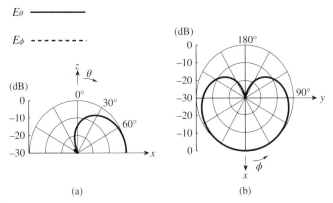

Figure 12.3 Radiation pattern for BeToL$_2$ at 3 GHz, where $(W_v, S_x, S_y) = (0.14\lambda_3, 0.22\lambda_3, 0.26\lambda_3,)$. (a) In the x–z plane. (b) In the x–y plane. (Reproduced from Ref. [1] with permission from IEEE.)

A parametric study is carried out at frequency f_3, where S_x and S_y are varied within the ranges $0.22\lambda_3 \leq S_x \leq 0.38\lambda_3$ and $0.22\lambda_3 \leq S_y \leq 0.34\lambda_3$, respectively, with the strip width W_v as a parameter. It is found that the BeToL radiates a wide unidirectional beam; representative BeToLs with parameters $(W_v, S_x, S_y) = (0.10\lambda_3, 0.22\lambda_3, 0.22\lambda_3)$, $(0.14\lambda_3, 0.22_{\lambda 3}, 0.26\lambda_3)$, and $(0.18\lambda_3, 0.22\lambda_3, 0.30\lambda_3)$ have a half-power beam width (HPBW) of greater than $150°$ in the H plane (x–y plane), with a small cross-polarized component (of less than -30 dB). For convenience, the BeToLs with these parameters are designated as BeToL$_1$, BeToL$_2$, and BeToL$_3$, respectively. Figure 12.3 shows the wide radiation pattern for BeToL$_2$.

The formation of the unidirectional radiation shown in Fig. 12.3 is explained using the currents on the conducting leaves. As seen from Fig. 12.4, a large current flows along the edge of each leaf. Figure 12.5 shows that phase-A (the phase of the current along the edge from a_b to a_u) is delayed relative to phase-B (the phase of

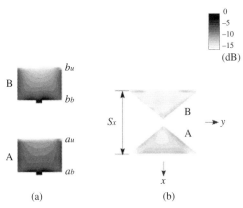

Figure 12.4 Current density for BeToL$_2$ at 3 GHz, with a ground plane of infinite extent. (a) Wide strip of leaves A and B. (b) Top triangular strip of leaves A and B. (Reproduced from Ref. [1] with permission from IEEE.)

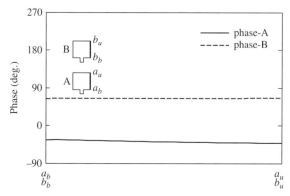

Figure 12.5   Phase of the currents along the edge from a_b to a_u on leaf A and the edge from b_b to b_u on leaf B. (Reproduced from Ref. [1] with permission from IEEE.)

the current from b_b to b_u). Due to this phase delay (approximately 90°) and a spacing of approximately one-quarter wavelength between the two leaves ($S_x \approx \lambda_3/4$), the radiation fields from leaves A and B add constructively in the positive x-space and destructively in the negative x-space. Note that the two vertical wide strips of the BeToL$_2$ behave like a two-monopole array (MPA), where both monopoles are excited with a phase difference of 90°; in other words, the HPBW for the BeToL$_2$ in the x–y plane is similar to the wide HPBW for the MPA (more than 150°, as calculated using the array factor [6]).

12.1.3   Input Impedance

The antenna dimensions of fed leaf A and shorted parasitic leaf B are the same. In addition, the distance between leaves A and B is relatively large ($S_x = 0.22$ wavelength at f_3 for the BeToL$_2$). In such a case, the input impedance for the BeToL is governed by fed leaf A, because there is no remarkable influence from parasitic leaf B, in contrast to the elements reported in Ref. 7. Figure 12.6 shows this fact, where

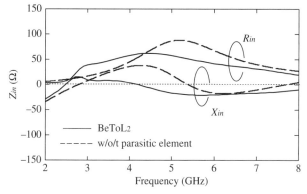

Figure 12.6   Input impedance for BeToL$_2$ in comparison with the input impedance for a fed leaf A in the absence of parasitic leaf B. (Reproduced from Ref. [1] with permission from IEEE.) Note $X_{in} < 0$ with $R_{in} > 0$ for BeToL$_2$ at frequencies below 2.5 GHz.

the input impedance of $BeToL_2$ (composed of fed leaf A and parasitic leaf B) and that of a fed leaf A in the absence of parasitic leaf B are presented. The $BeToL_2$ has a bandwidth of 86% (2.83–7.1 GHz) for a VSWR $= 2$ criterion (corresponding to a return loss criterion of RL $= -20\log|\Gamma| \approx 10$ dB, where $|\Gamma|$ is the voltage reflection coefficient [8]). Note that the VSWR bandwidth is 60% for $BeToL_1$ and 88% for $BeToL_3$.

12.1.4 Ground Plane

The ground plane for the BeToL in the previous sections is assumed to be of infinite extent. Further discussion is directed toward the effect of the ground plane size on the antenna characteristics, using $BeToL_2$. For this, the side length of the square ground plane is varied: $GP_x = GP_y = 1\lambda_3$, $2\lambda_3$, $3\lambda_3$, and $5\lambda_3$.

Figure 12.7 shows a comparison of the radiation patterns for $GP_x = GP_y = 3\lambda_3$ and $GP_x = GP_y = \infty$. The radiation is decomposed into two fields: one is the resultant field from leaves A and B and the other is the field radiated from the finite ground plane (GP-related field). The decomposition for $GP_x = GP_y = 3\lambda_3$ reveals that the maximum of the resultant field is on the positive x-axis and the maximum of the GP-related field is in a direction away from the horizontal plane. This means that the GP-related field shifts the maximum of the total radiation pattern upward in the x–z plane. The details for the frequency response of the direction of maximum radiation in the x–z plane, $(\theta, \phi) = (\theta_{max}, 0°)$, are shown in Fig. 12.8.

The gain in beam direction $(\theta, \phi) = (\theta_{max}, 0°)$ is a relatively high, as shown in Fig. 12.9. The gain for $GP_x = GP_y = 5\lambda_3$ is almost the same as that for $GP_x = GP_y = \infty$ and hence the curves for these two cases overlap. Note that, as the ground plane size is decreased, the radiation leaking toward the lower hemisphere (negative z-hemisphere) increases; this causes a change in the peak gain.

Figure 12.10 shows the frequency response of the return loss (RL), with the ground plane side length $GP_x = GP_y$ as a parameter. It is revealed that the return loss for a finite ground plane is very similar to that for a ground plane of infinite extent; as long as $GP_x = GP_y$ is greater than $1\lambda_3$, a frequency bandwidth of approximately 86% for an RL $= 10$ dB criterion is obtained.

The current density on the ground plane for $GP_x = GP_y = 3\lambda_3$ $(=300\,\text{mm})$ is presented in Fig. 12.11a. The current is found to concentrate on the positive x-side, coinciding with the beam direction. This implies that the area of the ground plane behind leaf B can be partially removed without affecting the antenna characteristics. Figure 12.11b illustrates the removal of the area $3\lambda_3 \times S_{GP/RMV}$.

12.1.5 Design Guidelines for The BeToL

Based on the investigation in Sections 12.1.2–12.1.4, the design guidelines for the BeToL are summarized as follows [1]. (1) Select the total leaf length, $S_z + t$, to be

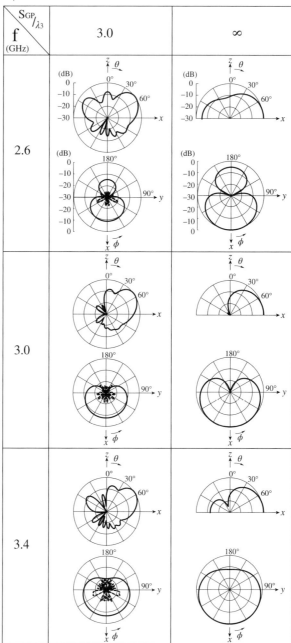

Figure 12.7 Comparison of the radiation patterns for two different ground planes: $GP_x = GP_y \equiv S_{GP} = 3\lambda_3$ (=300 mm) and $GP_x = GP_y \equiv S_{GP} = \infty$. (Reproduced from Ref. [1] with permission from IEEE.)

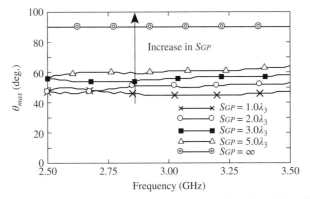

Figure 12.8 Frequency response of the direction of maximum radiation for BeToL$_2$, ($\theta = \theta_{max}$, $\phi = 0°$), with the side length of the square ground plane, $S_{GP} = GP_x = GP_y$, as a parameter. (Reproduced from Ref. [1] with permission from IEEE.)

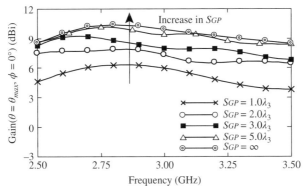

Figure 12.9 Frequency response of the gain for BeToL$_2$ in the direction of maximum radiation (θ_{max}, 0°) with the side length of the square ground plane, $S_{GP} = GP_x = GP_y$, as a parameter. (Reproduced from Ref. [1] with permission from IEEE.)

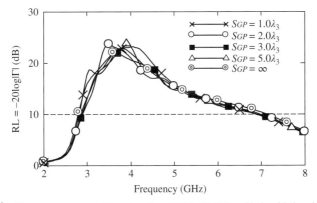

Figure 12.10 Frequency response of the return loss (RL) for BeToL$_2$ with the side length of the square ground plane, $S_{GP} = GP_x = GP_y$, as a parameter. (Reproduced from Ref. [1] with permission from IEEE.)

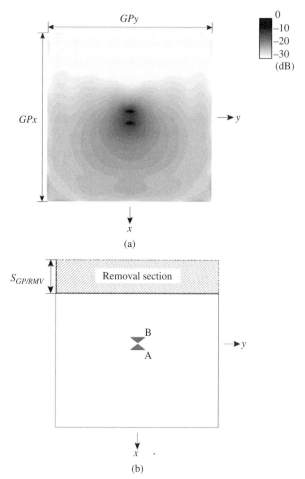

Figure 12.11 Current density and removed section for BeToL$_2$ at 3 GHz. (a) Current density for $GP_x = GP_y = 3\lambda_3$ (= 300 mm). (b) Removed section. (Reproduced from Ref. [1] with permission from IEEE.)

slightly smaller than one-quarter wavelength at the design frequency (e.g., $S_z + t = 0.22\lambda$), using $0.1\lambda \leq W_v \leq 0.18\lambda$ and $0.22\lambda \leq S_y \leq 0.30\lambda$. (2) Bend the leaf at S_z, which is approximately one-half the total leaf length (e.g., $S_z = 0.12\lambda$). (3) Fabricate a BeToL by arraying two leaves on the x-axis of the ground plane, where each leaf has the structure described in (1) and (2). (4) Excite the leaf on the positive x-axis, and short the leaf on the negative x-axis to the ground plane. Select the distance between the two leaves, S_x, to be approximately one-quarter wavelength, in order to form a wide H-plane radiation pattern in the positive x-space. (5) Fine-tune the distance S_x such that the phase difference of the current on the fed and parasitic leaves is approximately 90° out of phase (e.g., $S_x = 0.22\lambda$), using numerical methods (such as the FDTDM).

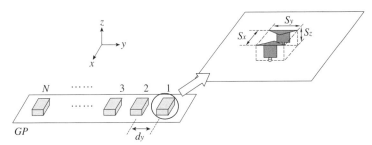

Figure 12.12 BeToL$_2$ array excited in-phase. (Reproduced from Ref. [9] with permission from IEEE.)

Note that the beam toward the positive x-space can be changed to a beam toward the negative x-space by shorting leaf A to the ground plane and feeding leaf B. This is realized using switching circuits with diodes or RF-MEMS. As long as the structure of the BeToL including the ground plane is symmetric with respect to the y-axis, the BeToL maintains the same antenna characteristics, but the direction will be switched, that is, the BeToL is reconfigurable.

12.1.6 BeToL Array

A reconfigurable antenna that has a high gain is realized by arraying BeToL$_2$ elements. Figure 12.12 shows such an array antenna [9], where the BeToL$_2$ elements are arrayed in the y-direction and excited in-phase. Figure 12.13 illustrates the gain as a function of the number of BeToL$_2$ elements, N, where the spacing between neighboring elements d_y is chosen to be less than one wavelength to avoid the appearance of grating lobes. As expected from array theory, the gain increases by approximately 3 dB, as the number of BeToL$_2$ elements is doubled [9]. Figure 12.14 shows the radiation patterns in the x–z and x–y planes at $N = 1, 2, 4,$

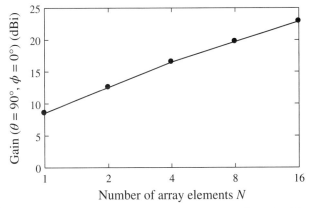

Figure 12.13 Gain as a function of the number of BeToL$_2$ elements. (Reproduced from Ref. [9] with permission from IEICE.)

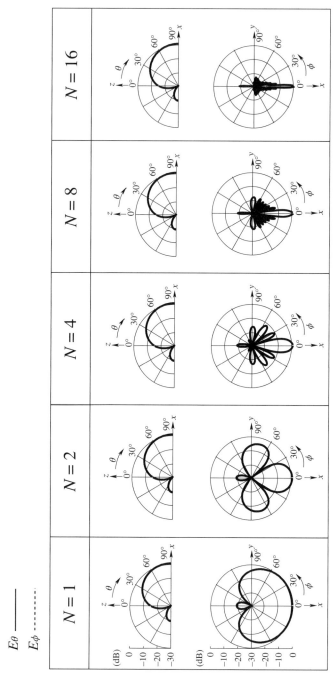

Figure 12.14 Radiation patterns. (Reproduced from Ref. [9] with permission from IEICE.)

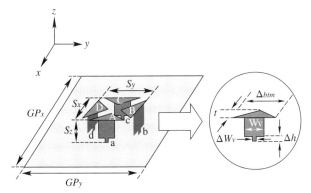

Figure 12.15 Reconfigurable BeFoL antenna. (Reproduced from Ref. [2] with permission from IEEE.)

8, and 16. A narrow half-power beam width (HPBW) is obtained in the x–y plane due to the array effect, while the HPBW in the x–z plane remains almost unchanged [9].

12.2 BeFoL ANTENNA

12.2.1 Configuration

The number of bent conducting leaves is increased to four, as shown in Fig. 12.15 [2]. This bent four-leaf antenna is abbreviated as the BeFoL, differentiating it from the BeToL. The bottom ends (a, c) of leaves (A, C) are on the x-axis, with a separation distance of S_x, and the bottom ends (b, d) of leaves (B, D) are on the y-axis with a separation distance of S_y. Distance S_x equals S_y. The top triangular leaf section is specified by a bottom length Δ_{btm} and height t. The configuration parameters used throughout this section are shown in Table 12.1.

Table 12.1 Configuration Parameters for the BeFoL

Symbol	Value
$\lambda_{4.5}$	66.7 mm
s_x	30 mm $= 0.45\lambda_{4.5}$
s_y	30 mm $= 0.45\lambda_{4.5}$
s_z	12 mm $= 0.18\lambda_{4.5}$
Δ_{btm}	20 mm $= 0.30\lambda_{4.5}$
t	10 mm $= 0.15\lambda_{4.5}$
w_v	14 mm $= 0.21\lambda_{4.5}$
Δw_v	2 mm $= 0.03\lambda_{4.5}$
Δh	1 mm $= 0.015\lambda_{4.5}$

12.2.2 Radiation Pattern

The situation where end a is excited, and ends b, c, and d are shorted to the ground plane is expressed as $[a]$. This produces a beam directed in the positive x-direction, due to the fact that the current along the vertical section of excited leaf A, which forms the E_θ component of the radiation pattern in the x–z plane, has a delayed phase with respect to the induced currents along the vertical sections of parasitic leaves B, C, and D. The radiation pattern for situation $[a]$ is illustrated at the bottom of Fig. 12.16. Note that the difference in orientation between adjacent leaves is 90°, hence the radiation is in the negative x-direction for situation $[c]$ (end c is excited and ends a, b, and d are shorted), in the positive y-direction for situation $[b]$ (end b is excited and ends a, c, and d are shorted), and in the negative y-direction for situation $[d]$ (end d is excited and ends a, b, and c are shorted). Thus, a reconfigurable radiation pattern is obtained by specifying how the leaf ends are connected.

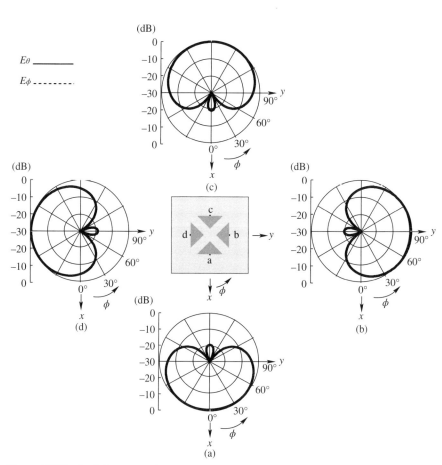

Figure 12.16   Radiation patterns in the azimuth plane (x–y plane) at a test frequency of 4.5 GHz. (Reproduced from Ref. [2] with permission from IEEE.) (a) Situation [a]. (b) Situation [b]. (c) Situation [c]. (d) Situation [d].

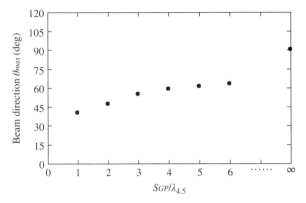

Figure 12.17 Beam direction for situation [a] in the elevation plane (x–z plane) as a function of the ground plane side length $S_{GP} = GP_x$ ($= GP_y$) at a frequency of 4.5 GHz. (Reproduced from Ref. [2] with permission from IEEE.)

12.2.3 Bent Ground Plane

A beam in or near the horizontal direction is often required in practical applications, under the condition that the ground plane size is finite. Figure 12.17 shows the beam direction θ_{max} (measured from the z-axis) for situation [a] as a function of the ground plane side length GP_x ($= GP_y$) normalized to the wavelength at 4.5 GHz. As GP_x is decreased from GP_x ($= GP_y$) $= \infty$, the maximum beam direction deviates from the horizontal direction. One technique to realize horizontal radiation is to bend the ground plane at $x = \pm b_{GP}$ and $y = \pm b_{GP}$ (b_{GP} is called the *bend line coordinate*), as shown in Fig. 12.18, thereby controlling the contribution of the radiation from the induced

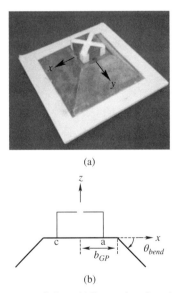

Figure 12.18 BeFoL on a bent ground plane. (a) Perspective view photo. (b) Side view. (Reproduced from Ref. [2] with permission from IEEE.)

current flowing on the ground plane to the total radiation. Note that the bend angle with respect to the horizontal plane (x–y plane) is expressed as θ_{bend}.

Figure 12.19 shows the radiation pattern for situation [a] when the bend angle θ_{bend} is varied, where the bend line coordinate is $b_{GP} = 20\Delta = 0.3\lambda_{4.5}$ and the ground plane side length is $GP_x = GP_y = 124\Delta = 1.86\lambda_{4.5}$ ($\Delta = 1$ mm). Note that b_{GP} is chosen to be small in order to increase the contribution of the radiation from the part of the ground plane beyond the bend line coordinate to the total radiation. It is found that there is an optimum angle of θ_{bend} for realizing confined radiation around the positive x-axis. This confined radiation leads to a high antenna gain in the positive x-direction, compared with the gain for the flat ground plane, as shown in Fig. 12.20. The gain before and after bending the ground plane is 2.9 dBi and 6.2 dBi (for $\theta_{bend} = 45°$), respectively, in the positive x-direction at 4.5 GHz. Figure 12.21 shows the VSWR frequency response for $\theta_{bend} = 45°$. The VSWR at frequencies around 4.5 GHz is less than 2, as desired.

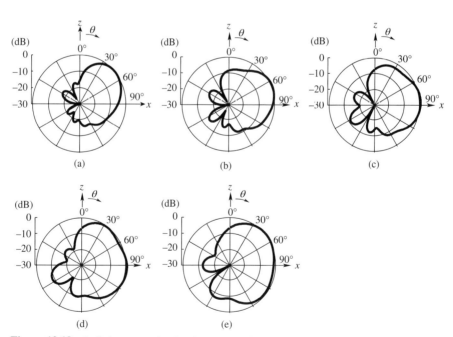

Figure 12.19 Radiation pattern for situation [a] at 4.5 GHz, where $(GP_x, b_{GP}) = (124\Delta,$ $20\Delta) = (1.86\lambda_{4.5}, 0.3\lambda_{4.5})$.(a) $\theta_{bend} = 0°$ (flat ground plane). (b) $\theta_{bend} = 26.6°$. (c) $\theta_{bend} = 45°$. (d) $\theta_{bend} = 63.4°$. (e) $\theta_{bend} = 90°$. (Reproduced from Ref. [2] with permission from IEEE.)

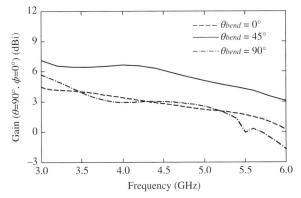

Figure 12.20 Frequency response of the gain for situation [a] in the horizontal direction, where (GP_x, b_{GP}) = (124Δ, 20Δ) = (1.86$\lambda_{4.5}$, 0.3$\lambda_{4.5}$). (Reproduced from Ref. [2] with permission from IEEE.)

Exercise

BeToL, BeFoL, and antennas in Refs 10–14 can be reconfigured by selecting (changing) the feed point. Design a reconfigurable antenna that has a single fixed feed point.

Answer Figure 12.22 shows a design example [15], which consists of a square patch (of side length S_{out} = 0.24 wavelength) with a slot and four parasitic bent Y elements (Y_H = 0.09 wavelength). The patch is located at a height of h = 0.08 wavelength above the ground plane (GP) and excited from point F (the origin of the rectangular coordinates). Inverted-L lines near the feed point are used for impedance matching. The bent Y elements are designed such that their bottom ends a, b, c, and d can be selected to be open or shorted to the ground plane

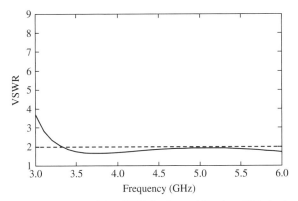

Figure 12.21 Frequency response of the VSWR for θ_{bend} = 45°, where (GP_x, b_{GP}) = (124Δ, 20Δ) = (1.86$\lambda_{4.5}$, 0.3$\lambda_{4.5}$). (Reproduced from Ref. [2] with permission from IEEE.)

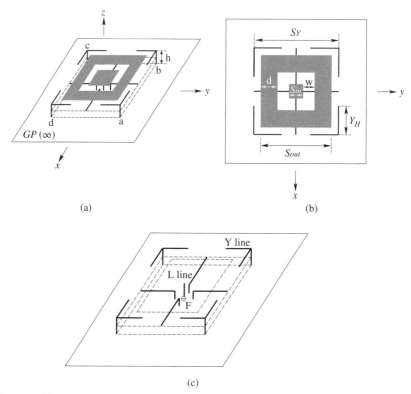

(a)   (b)

(c)

Figure 12.22 Reconfigurable antenna with a fixed feed point. (a) Perspective view. (b) Top view. (c) View under the patch. (Reproduced from Ref. [15] with permission from IEEE.)

by switching circuits. Table 12.2 shows the state of the bottom ends (cases), where o and s denote *open* and *shorted*, respectively. The radiation patterns for cases 1, 2, 3, and 4 at 2.4 GHz (wavelength $\lambda_{2.4} = 12.5$ cm) are illustrated in Fig. 12.23, where the parameters in Table 12.3 are used. The photo of a commercialized sample is shown in Fig. 12.24 (courtesy of Nihon Dengyo Kosaku Co.). ∎

Table 12.2 State of the Bottom Ends

	a	b	c	d
Case 1	o	s	s	o
Case 2	o	o	s	o
Case 3	o	o	s	s
Case 4	o	o	o	s
Case 5	s	o	o	s
Case 6	s	o	o	o
Case 7	s	s	o	o
Case 8	o	s	o	o

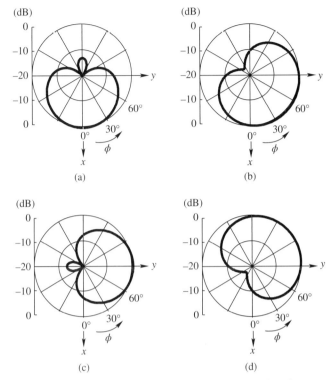

Figure 12.23 Radiation patterns. (a) Case 1. (b) Case 2. (c) Case 3. (d) Case 4.

Table 12.3 Parameters

Symbol	Value
f	2.4 GHz
S_{out}	$0.24\lambda_{2.4}$
S_{in}	$0.06\lambda_{2.4}$
w	$0.03\lambda_{2.4}$
d	$0.06\lambda_{2.4}$
S_y	$0.28\lambda_{2.4}$
Y_H	$0.09\lambda_{2.4}$
h	$0.08\lambda_{2.4}$

Figure 12.24   Production antenna sample, where the same design concept as that presented in Ref. 15 is used. (Courtesy of Nihon Dengyo Kosaku Co.)

REFERENCES

1. H. Nakano, Y. Ogino, and J. Yamauchi, Bent two-leaf antenna radiating a tilted, linearly polarized, wide beam. *IEEE Trans. Antennas Propag.*, vol. 58, no. 11, pp. 3721–3725, 2010.
2. H. Nakano, Y. Ogino, and J. Yamauchi, Bent four-leaf antenna. *IEEE Antennas Wireless Propag. Lett.*, vol. 10, pp. 223–226, 2010.
3. S. L. Preston, D. V. Thiel, J. W. Lu, S. G. O'Keefe, and T. S. Bird, Electronic beam steering using switched parasitic patch elements. *Electron. Lett.*, vol. 33, no. 1, pp. 7–8, 1997.
4. M. R. Kamarudin, P. S. Hall, F. Colombel, and M. Himdi, Switchable disk-loaded monopole antenna array with CPW feeding system. *Proc. IEEE AP-S Int. Symp.*, Albuquerque, NM, July 2006, pp. 2313–2316.
5. P. Deo, A. Mehta, D. Mirshekar-Syahkal, P. J. Massey, and H. Nakano, Thickness reduction and performance enhancement of steerable square loop antenna using hybrid high impedance surface. *IEEE Trans. Antennas Propag.*, vol. 58, no. 5, pp. 1477–1485, 2010.
6. W. L. Stutzman and G. A. Thiele, *Antenna Theory and Design*, chapter 3, New York: Wiley, 1981.
7. H. Nakano, R. Suzuki, and J. Yamauchi, Low-profile inverted-F antenna with parasitic elements on an infinite ground plane. *IEE Proc. Microw. Antennas Propag.*, vol. 145, no. 4, pp. 321–325, 1998.
8. D. M. Pozar, *Microwave Engineering*, 2nd edn, New York: Wiley, 1998, pp. 67–67.
9. Y. Sato, J. Yamauchi, and H. Nakano, Radiation characteristics of a bent two-leaf array antenna. *Proc. IEICE Society conference*, B-1-116, Sendai, 2015.
10. A. Mehta, D. Mirshekar-Syahkal, and H. Nakano, Beam adaptive single arm rectangular spiral antenna with switches. *IEE Proc. Microw. Antennas Propag.*, vol. 153, no. 1, pp. 13–18, 2006.
11. A. Pal, A. Mehta, D. Mirshekar-Syahkal, and P. J. Massey, Doughnut and tilted beam generation using a single printed star antenna. *IEEE Trans. Antennas Propag.*, vol. 57, no. 10, pp. 3413–3418, 2009.
12. A. Pal, A. Mehta, D. Mirshekar-Syahkal, and H. Nakano, A square-loop antenna with 4-Port feeding network generating semi-doughnut pattern for vehicular and wireless applications. *IEEE Antennas Wirel. Propag. Lett.*, vol. 10, pp. 338–341, 2011.

13. P. Deo, A. Mehta, D. Mirshekar-Syahkal, P. J. Massey, and H. Nakano, Thickness reduction and performance enhancement of steerable square loop antenna using hybrid high impedance surface. *IEEE Trans. Antennas Propag.*, vol. 58, no. 5, pp. 1477–1485, 2010.

14. A. Pal, A. Mehta, D. Mirshekar-Syahkal, P. Deo, and H. Nakano, Dual-band low-Profile capacitively coupled beam-steerable square-loop antenna. *IEEE Trans. Antennas Propag.*, vol. 62, no. 3, pp. 1204–1211, 2014.

15. H. Nakano, R. Aoki, R. Kobayashi, and J. Yamauchi, A patch antenna surrounded by parasitic Y elements for beam scanning. *Proc. IEEE AP-S Int. Symp.*, July 2006, pp. 2317–2320.

Chapter 13

Patch Antenna with a Nonuniform Loop Plate

13.1 ANTENNA SYSTEM

Figure 13.1a shows an antenna system consisting of a fed antenna and a parasitic dielectric plate that has periodically arrayed elements (PerioAEs), such as conducting patches, crosses, or loops. Due to the Fabry–Pérot (FabPe) phenomenon [1–3], the antenna system radiates a narrow beam in the direction normal to the PerioAEs plate, that is, in the z-direction. When a radiation beam in a direction θ, measured from the z-axis, is required the entire antenna system must be tilted by angle θ, as shown in Fig. 13.1b. However, if the radiation beam from the antenna system is inherently tilted with angle θ, as shown in Fig. 13.1c, then it is not necessary to tilt the antenna system. This antenna system in Fig. 13.1c has a lower profile than the antenna system in Fig. 13.1b, and hence it has a broader field of application [4].

Figure 13.2 shows the antenna structure to be considered in this chapter, which is based on the FabPe resonator concept, where a patch printed on substrate 1 is used as a fed antenna and a PerioAEs plate (where PerioAEs are printed on the reverse side of substrate 2 is placed above the fed patch antenna with spacing d_{12}. The patch is square with area $l_{patch} \times l_{patch}$, and the feed point is displaced from the center point by distance x_F. Substrate 1 is specified by area $s_1 \times s_1$, relative permittivity ε_{r1}, and thickness B_1, while substrate 2 is specified by area $s_x \times s_y$, relative permittivity ε_{r2}, and thickness B_2. The ground plane under substrate 1 has area $s_{GP} \times s_{GP}$. Note that the side length s_x is chosen to be equal s_1 ($s_x = s_1$). Also, note that the center points of the PerioAEs plate, square patch, substrate 1, and square ground plane are all located at $(x, y) = (0, 0)$.

PerioAEs consisting of numerous square strip loops are shown in Fig. 13.2b, whose inner and outer side lengths are s_m and s_{out}, respectively, as shown in Fig. 13.2c. All loops have the same outer loop side length s_{out} and the spacing

Low-Profile Natural and Metamaterial Antennas: Analysis Methods and Applications, First Edition. Hisamatsu Nakano.
© 2016 The Institute of Electrical and Electronics Engineers, Inc. Published 2016 by John Wiley & Sons, Inc.

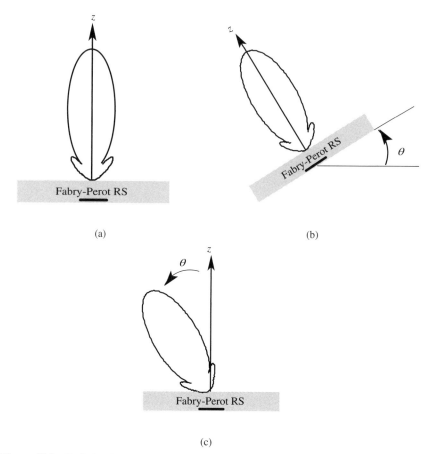

Figure 13.1   Radiation from a Fabry–Perot (FabPe) resonator. (a) Antenna with broadside radiation. (b) Physical/mechanical rotation of the antenna shown in (a). (c) Antenna with a tilted beam. (Reproduced from Ref. [4] with permission from IEEE.)

between neighboring loops, Δs, is constant. The width of the loop for row number m ($= 1, 2, \ldots, M$), w_m, is

$$w_m = (s_{out} - s_m)/2 \tag{13.1}$$

The side lengths of the PerioAEs plate are

$$s_x = s_1 = M(s_{out} + \Delta s) \tag{13.2}$$

$$s_y = (2N - 1)(s_{out} + \Delta s) \tag{13.3}$$

where M and N are integers indicating the total number of loops: $M \times (2N - 1)$.

The parameters shown in Tables 13.1 and 13.2 are fixed throughout this chapter, and designated as the fundamental parameters for the following discussions. The side length of the ground plane, s_{GP}, number of loops in the y-direction, $2N - 1$ (and hence the area of substrate 2), and strip width of the loops, w_m, are not

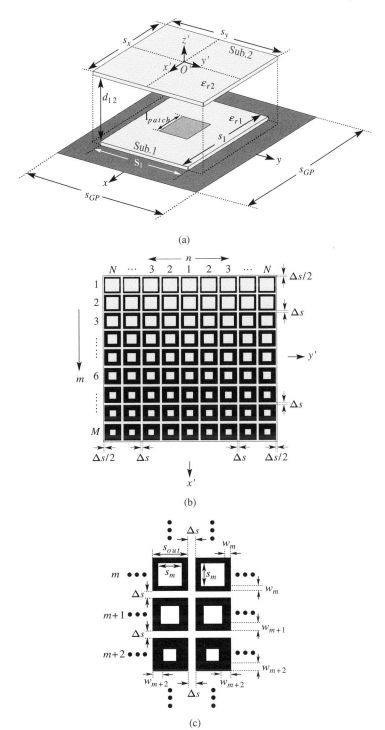

Figure 13.2 Antenna system. (a) Perspective view. (b) Square loops printed on the reverse side of substrate 2. (c) Neighboring loops. (Reproduced from Ref. [4] with permission from IEEE.)

Table 13.1   Parameters for Patch and Substrate 1

Symbol	Value	Symbol	Value
l_{patch}	10.5 mm	ε_{r1}	2.6
x_F	2.5 mm	s_1	135 mm
B_1	1 mm		

Table 13.2   Parameters for Loop and Substrate 2

Symbol	Value	Symbol	Value
s_{out}	13. 5 mm	ε_{r2}	2.2
Δs	1.5 mm	d_{12}	18 mm
M	9	s_x	135 mm
B_2	1 mm		

fixed and are varied according to the objectives of the analysis. These nonfixed parameters are expressed by the set (s_{GP}, $2N-1$, w_m), and are designated as the object parameters.

13.2 REFERENCE GAIN AND BROADSIDE RADIATION— PLACEMENT OF A HOMOGENENEOUS PerioAEs PLATE

Figure 13.3 shows an isolated fed patch antenna with the parameters shown in Table 13.1, which is used for the following discussion, where $s_{GP} \times s_{GP} = s_1 \times s_1$. As shown in Fig. 13.4, the patch radiates a unidirectional wide beam. This radiation pattern is calculated using the FDTDM, for which the side length of a unit cell is chosen to be Δ_{FDTD} (= 0.25 mm). The gain in the z-direction at a test frequency of 8 GHz is 5.5 dBi, and is used as the reference gain.

Using the patch in Fig. 13.3, a PerioAEs plate (substrate 2) is located above the patch and the patch is excited. Substrate 2 has $M \times (2N-1) = 9 \times 9$ printed loops, and hence, the side lengths of substrate 2 are $s_x = s_y = 135$ mm, as calculated from

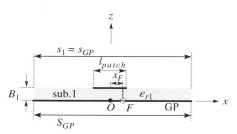

Figure 13.3   Side view of the reference patch, where the ground plane GP and substrate1 have the same size ($s_{GP} \times s_{GP} = s_1 \times s_1$). (Reproduced from Ref. [4] with permission from IEEE.)

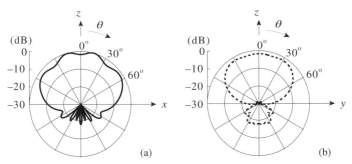

Figure 13.4 Radiation from the reference patch at 8 GHz. (a) In the x–z plane. (b) In the y–z plane. (Reproduced from Ref. [4] with permission from IEEE.)

Eqs. (13.2) and (13.3). All the inner side lengths of the strip loops are chosen to be the same (homogeneous): $s_m = 8$ mm, resulting in a uniform loop width of $w_m = 2.75$ mm ($m = 1, 2, \ldots, 9$). The other parameters for the loops and substrate 2 are defined in Table 13.2 as the fundamental parameters. The object parameters are summarized as $(s_{GP}, 2N - 1, w_m) = (135\ \text{mm}, 9, 2.75\ \text{mm})$.

Investigation of the aperture field distribution over a horizontal plane just above the top surface of substrate 2 (at $z' = 0.25$ mm) reveals that there is a wide area where the x'-directed electric fields (aperture fields) are almost equal in amplitude with respect to the x'-axis and are in-phase. Consequently, the θ-directed radiation field E_θ in the x–z plane and the ϕ-directed radiation field E_ϕ in the y–z plane form narrow patterns, as shown in Fig. 13.5. The HPBW is 14° in the x–z plane and

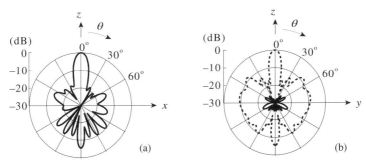

Figure 13.5 Radiation pattern at 8 GHz. (a) In the x–z plane. (b) In the y–z plane. A homogeneous PerioAEs plate (substrate 2) is placed above the fed patch. The parameters $(s_{GP}, 2N - 1, w_m) = (135\ \text{mm}, 9, 2.75\ \text{mm})$ and the parameters in Tables 13.1 and 13.2 are used. (Reproduced from Ref. [4] with permission from IEEE.)

12° in the y–z plane. The gain is increased by 11.3 dB from the reference gain, due to the constructive effect of the x'-directed electric field distribution over the top surface of substrate 2. Note that a radiation field E_ϕ in the x–z plane (cross-polarization component) does not appear due to the fact that the y'-directed aperture electric fields are almost equal in amplitude with respect to the x'-axis and are out-of-phase.

13.3 GRADATION CONSTANT AND TILTED RADIATION BEAM—PLACEMENT OF A NONHOMOGENEOUS PerioAEs PLATE

A homogeneous PerioAEs plate has an in-phase electric field distribution and forms a narrow radiation beam in the z-direction with a high gain. In this subsection, the in-phase electric field distribution is changed in order to form a tilted beam. The change is made by gradating the strip widths of the loops, specified using a gradation constant $\Delta w\ (= w_{m+1} - w_m)$, that is, the difference between the strip widths of neighboring loops in the x'-direction. Thus, the homogeneous PerioAEs plate is changed to a nonhomogeneous PerioAEs.

Table 13.3 shows three gradation examples, where the sixth strip loop width w_6 is chosen as the reference width (2.75 mm $= 11\Delta_{\mathrm{FDTD}}$) and the mth strip loop width w_m is defined as

$$w_m = w_6 + i\Delta w,$$
$$i = m - 6\ (m = 1, 2, \ldots, M) \tag{13.4}$$

Note that the loop widths w_1 and w_2 for a gradation constant of $\Delta w = 0.75$ mm $= 3\Delta_{\mathrm{FDTD}}$ are zero, that is, loops for $m = 1$ and 2 do not exist.

The radiation patterns for these gradations at 8 GHz are shown in Fig. 13.6, where the object parameters $(s_{GP}, 2N - 1, w_m) = (135$ mm, 9, varied) and the fundamental parameters in Tables 13.1 and 13.2 are used. It is found that the nonhomogeneous PerioAEs plate forms a tilted beam in the negative x-space.

Table 13.3   Loop width w_m and gradation constant Δw

w_m	$\Delta w = 0.25$ mm	$\Delta w = 0.50$ mm	$\Delta w = 0.75$ mm
w_1	1.50 mm	0.25 mm	
w_2	1.75 mm	0.75 mm	
w_3	2.00 mm	1.25 mm	0.50 mm
w_4	2.25 mm	1.75 mm	1.25 mm
w_5	2.50 mm	2.25 mm	2.00 mm
w_6	2.75 mm	2.75 mm	2.75 mm
w_7	3.00 mm	3.25 mm	3.50 mm
w_8	3.25 mm	3.75 mm	4.25 mm
w_9	3.50 mm	4.25 mm	5.00 mm

Source: Reproduced from Ref. 4 with permission from IEEE.

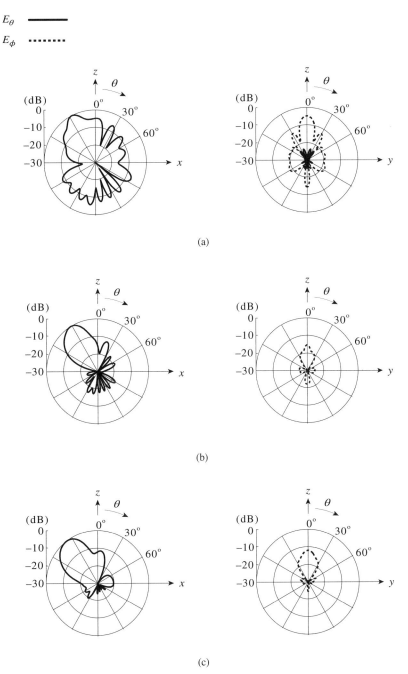

E_θ ———

E_ϕ ········

(a)

(b)

(c)

Figure 13.6 Tilted radiation patterns at 8 GHz, where the gradation constant Δw is (a) $\Delta w = 0.25$ mm, (b) $\Delta w = 0.5$ mm, and (c) $\Delta w = 0.75$ mm. (Reproduced from Ref. [4] with permission from IEEE.) The parameters $(s_{GP}, 2N-1, w_m) = (135$ mm, 9, varied) and the parameters in Tables 13.1 and 13.2 are used.

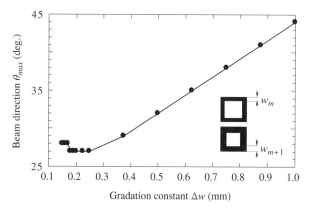

Figure 13.7   Beam tilt angle (beam direction) θ_{max} as a function of gradation constant Δw. The parameters $(s_{GP}, 2N-1, w_m) = (135\,\text{mm}, 9, \text{varied})$ and the parameters in Tables 13.1 and 13.2 are used. (Reproduced from Ref. [4] with permission from IEEE.)

The radiation patterns in Fig. 13.6 are better understood from the aperture electric fields E_x and E_y, that is, the fields over a horizontal plane just above substrate 2. It is found that the x'-directed aperture field E_x is symmetric with respect to the x'-axis and the phase $\angle E_x$ lags toward the negative x'-direction. This forms the maximum radiation field E_θ within the negative x-space of the x–z plane, according to array theory [5]. Note that the radiation field component E_ϕ in the x–z plane (cross-polarization component) does not appear due to the fact that the y'-directed aperture field E_y is asymmetric with respect to the x'-axis (equal amplitude and out-of-phase distributions). Figure 13.7 shows the details of the relationship between the gradation constant Δw and the beam tilt angle (beam direction) θ_{max}.

13.4   GAIN

The beam direction θ_{max} is controlled by the gradation constant Δw. For investigation of the gain in the beam direction θ_{max}, two parameters, α^2 and β^2 (area ratios), are introduced for a square ground plane of side length s_{GP}.

$$\alpha^2 \equiv (s_x \times s_y)/s_{GP}^2 \tag{13.5}$$

$$\beta^2 \equiv s_1^2/s_{GP}^2 \tag{13.6}$$

The gain depends on the in-phase area spreading out over the horizontal plane just above substrate 2. Figure 13.8a illustrates the gain (the absolute gain is shown, and hence the loss due to impedance mismatch is not included) for a gradation constant of $\Delta w = 0.5\,\text{mm}$ at a frequency of 8 GHz, where the number of loops in the y'-direction, $2N-1$, is varied, that is, the side length in the

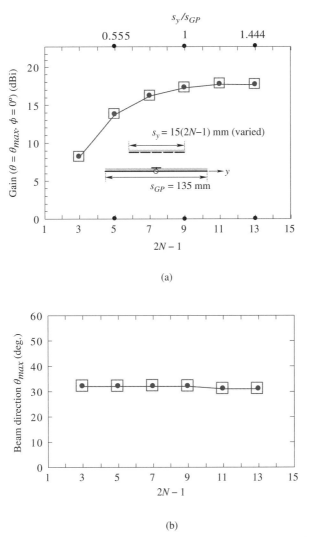

Figure 13.8 Gain in the beam direction at 8 GHz, when the number of loops in the y'-direction, $2N-1$, is varied with $\beta^2 = 1$. (a) Gain. (Reproduced from Ref. [4] with permission from IEEE.) (b) Beam direction. The parameters $(s_{GP}, 2N-1, w_m) = (135\,\text{mm},$ varied, varied with $\Delta w = 0.5\,\text{mm})$ and the parameters in Tables 13.1 and 13.2 are used.

y'-direction, s_y, is varied subject to Eq. (13.3) [note that the number of loops in the x'-direction, M, is fixed at $M = 9$, as already specified in Table 13.2]. The following parameters are used: side lengths $s_x = s_1$ and $s_{GP} = s_1$ [=135 mm, as shown in Table 13.1, leading to $\alpha^2 = s_y/s_{GP} = (2N-1)(s_{out} + \Delta s)/s_{GP}$ and $\beta^2 = 1$]. It is revealed that the gain in the beam direction [$\theta = \theta_{max}$ shown in Fig. 13.8b] increases with an increasing number of loops, $2N-1$, and levels off at $2N-1 \approx 9$, and that the beam direction is less sensitive when $2N-1 \geq 3$.

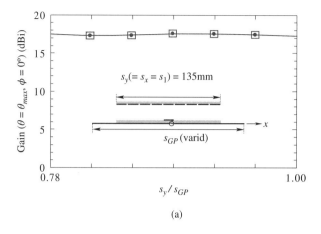

(a)

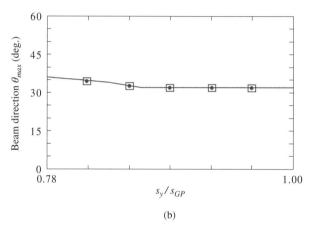

(b)

Figure 13.9   Frequency response of the gain at 8 GHz as a function of the side length of the square ground plane, s_{GP}. (a) Gain. (b) Beam direction. (Reproduced from Ref. [4] with permission from IEEE.) The parameters $(s_{GP}, 2N-1, w_m) = $ (varied, 9, varied with $\Delta w = 0.5$ mm) and $(s_x, s_y) = (s_1, s_1)$ are used, together with the parameters in Tables 13.1 and 13.2. Note that $\alpha = \beta = s_y/s_{GP} = s_1/s_{GP}$.

Next, the investigation is directed to the gain when the area of the square ground plane, $s_{GP} \times s_{GP}$, is varied. Figure 13.9a shows the gain at 8 GHz with s_{GP} as a parameter, where $(s_{GP}, 2N-1, w_m) = $ (varied, 9, varied with $\Delta w = 0.5$ mm) and $(s_x, s_y) = (s_1, s_1)$ [$s_1 = 135$ mm, as specified in Table 13.1, leading to $\alpha^2 = \beta^2 = (s_y/s_{GP})^2 - (s_1/s_{GP})^2$] are used. It is found that variation in the gain due to the ground plane size is not remarkable. Note that the gain in Fig. 13.9a is calculated in the beam direction θ_{max}, which is shown in Fig. 13.9b.

Up to this point, the gain is investigated under the condition that substrate 2 is lossless ($\tan\delta = 0$). When substrate 2 is lossy, it causes a decrease in the gain. Figure 13.10 shows the frequency response of the gain with $\tan\delta$ as a parameter,

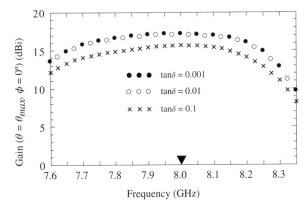

Figure 13.10 Frequency response of the gain with tanδ as a parameter. (Reproduced from Ref. [4] with permission from IEEE.)

where $(s_{GP}, 2N-1, w_m) = (135 \text{ mm}, 9, \text{ varied with } \Delta w = 0.5 \text{ mm})$ and the parameters in Tables 13.1 and 13.2 are used. The investigation reveals that the gain reduction is very small as long as tanδ is less than 0.01.

REFERENCES

1. G. V. Trentini, Partially reflecting sheet arrays. *IRE Trans. Antennas Propag.*, vol. AP-4, pp. 666–671, 1956.
2. A. Feresidis, G. Goussetis, S. Wang, and J. Vardaxoglou, Artificial magnetic conductor surfaces and their application to low-profile high-gain planar antennas. *IEEE Trans. Antennas Propag.*, vol. 53, no. 1, pp. 209–215, 2005.
3. N. Guérin, S. Enoch, G. Tayeb, P. Sabouroux, P. Vincent, and H. Legay, A metallic Fabry–Perot directive antenna. *IEEE Trans. Antennas Propag.*, vol. 54, no. 1, pp. 220–224, 2006.
4. H. Nakano, S. Mitsui, and J. Yamauchi, Tilted-beam high gain antenna system composed of a patch antenna and periodically arrayed loops. *IEEE Trans. Antennas Propag.*, vol. 62, no. 6, pp. 2917–2925, 2014.
5. W. Stutzman and G. Thiele, *Antenna theory and design*, NY: Wiley, 1981.

Chapter 14

Linearly Polarized Rhombic Grid Array Antenna

A conventional grid array antenna (GAA [1,2]) is composed of wire radiation elements and wire feed elements, both backed by a conducting plate (ground plane). Kraus, the inventor of the GAA, qualitatively explained the radiation mechanism for backward radiation, assuming a traveling wave current along the wires of the radiation elements and feed elements. Nakano et al. were the first to confirm this assumption of a traveling wave current [3], using the method of moments [4].

The wires constituting the GAA [5] can be replaced by equivalent strip lines [6–10]. In Ref. 6, the antenna characteristics for a strip GAA are compared with those for a wire GAA. In Refs 7 and 8, the antenna characteristics when the space between the strip grid array and the ground plane (SPACE$_{GAA-GP}$) is filled with a rigid dielectric material are investigated; in Refs 9 and 10, the antenna characteristics when the SPACE$_{GAA-GP}$ is filled with a composite of two layers (a solid dielectric layer and an air layer) are discussed. Note that these strip GAAs [6–10] are designed to have maximum radiation in the direction normal to the antenna plane, that is, they have a broadside beam. Such a broadside beam is also found in Refs. 11 and 12.

This chapter presents a novel GAA [13] from the perspective of realizing a beam-scanning antenna. The radiation elements for this GAA are bent with angle 2α (see Fig. 14.1), forming a structure with numerous rhombic cells. This GAA is referred to as a *rhombic grid array antenna* (R-GAA). Note that when $2\alpha = 180°$ the antenna corresponds to a conventional Kraus GAA [1,2]. The reader will find that the gain for the R-GAA has less variation with frequency than the gain for the conventional GAA over the band of interest, as desired for beam scanning.

14.1 CONFIGURATION

Figure 14.1 shows the R-GAA. Both the x-directed elements and y-directed elements lie in the same plane at a very low height (relative to the operating

Low-Profile Natural and Metamaterial Antennas: Analysis Methods and Applications, First Edition.
Hisamatsu Nakano.

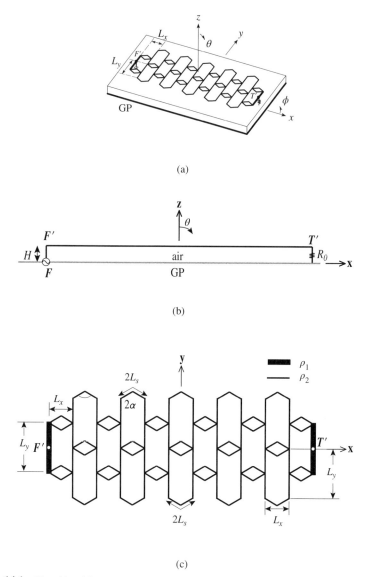

(a)

(b)

(c)

Figure 14.1   Rhombic grid array antenna. (a) Perspective view. (b) Side view. (c) Top view. (Reproduced from Ref. [13] with permission from IEEE.)

wavelength) above a conducting ground plane (GP). F', the starting point of the array, is located at the left edge, to which the inner conductor of a coaxial feed line is connected. The outer conductor of the coaxial feed line is soldered to the ground plane. T', the ending point of the array, is at the right edge and is shorted to the ground plane through a resistor R_0.

14.2 RADIATION PATTERN AND GAIN

Analysis is performed using the method of moments [4], where the ground plane, GP, is assumed to be of infinite extent. The antenna configuration parameters that remain unchanged throughout this chapter are listed in Table 14.1. Note that, at 6 GHz, a spacing of $L_x = 25$ mm corresponds to one-half wavelength ($0.5\lambda_6$, where λ_6 is the free-space wavelength at 6 GHz), and a length of $L_y = 50$ mm corresponds to one wavelength ($1\lambda_6$). The bend angle 2α is varied subject to the objectives of the analysis.

The x-directed bent elements act as radiation elements and the y-directed straight elements act as feed elements for the radiation elements. The radiation beam is directed in the z-direction (normal to the grid array) when all the radiation elements are excited in-phase. The frequency for this condition is denoted as $f_{in\text{-}phase}$. As the frequency f increases or decreases from $f_{in\text{-}phase}$, the excitation deviates from the in-phase condition and the radiation pattern in the x–z plane (elevation plane) becomes tilted, as shown in Fig. 14.2, where E_θ and E_ϕ denote the θ and ϕ components of the radiation field, respectively.

Details for when the beam direction $\theta = \theta_{max}$ as a function of frequency f are shown in Fig. 14.3a, where three bend angles are used: $2\alpha = 90°$, $120°$, and $180°$. The angle θ_{max} is measured from the z-axis in the x–z plane (elevation plane); a positive θ_{max} indicates forward radiation (toward the positive x-space) and a negative θ_{max} indicates backward radiation (toward the negative x-space). As the bend angle 2α is decreased, the frequency response curve shifts to the left relative to that of a conventional GAA structure ($2\alpha = 180°$). This results from an increase in the radiation element length $2L_s$ ($= L_x/\sin\alpha$) under a fixed L_x. Note that the experimental data in Fig. 14.3a (as well as in Figs. 14.4 and 14.6) were obtained using a fabricated antenna, shown in Fig. 14.3b.

Figure 14.4 shows the frequency response of the gain in the beam direction θ_{max}. It is found that, across a frequency range of f_3 (near $f_3^{2\alpha}$) to f_4 (near $f_4^{2\alpha}$), the gain for the R-GAA with bent radiation elements has less variation than that for a

Table 14.1 Configuration Parameters

Symbol	Value	Unit
L_x	25	mm
L_y	50	mm
$2L_s$ ($2\alpha = 90°$)	35.36	mm
$2L_s$ ($2\alpha = 120°$)	28.87	mm
$2L_s$ ($2\alpha = 180°$)	25	mm
ρ_1	0.7	mm
ρ_2	0.2	mm
H	2.5	mm

Source: Reproduced from Ref. 13 with permission from IEEE.

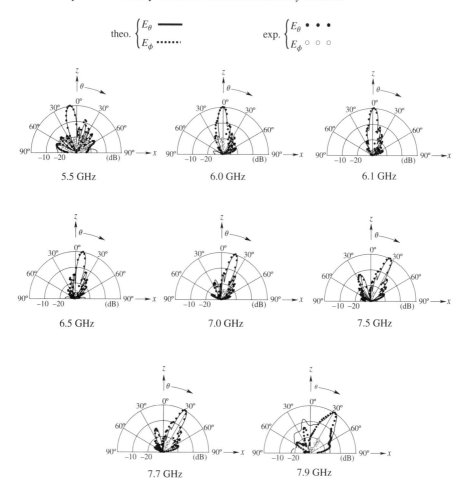

Figure 14.2   Radiation pattern for the R-GAA, where $2\alpha = 120°$. (Reproduced from Ref. [13] with permission from IEEE.)

conventional GAA with straight radiation elements ($2\alpha = 180°$), as desired for beam scanning. Note that the notation $f_p^{2\alpha}$ ($p = 3$, 4) denotes the frequencies when the perimeter of the closed loop (the bold dotted line shown at the top of Fig. 14.5) corresponds to p-guided wavelengths ($p\lambda_g$) for bend angle 2α. The variation in the gain, ΔG from frequency f_3 to f_4 is 1.6 dB for $2\alpha = 90°$, 0.6 dB for $2\alpha = 120°$, and 3.0 dB for $2\alpha = 180°$.

Another finding in Fig. 14.4 is that the gain decreases singularly at frequencies $f_3^{2\alpha}$ and $f_4^{2\alpha}$. This can be explained using the current distribution. Looking at Fig. 14.5a and b, where the current distribution for a representative bend angle $2\alpha = 120°$ at $f_4^{120} = 7.9$ GHz (let us call this a singular gain frequency) and 7 GHz (a nonsingular gain frequency) are presented, respectively. The current distribution is symmetric with respect to the x-axis, and hence, only the current distribution on the positive y-side is illustrated. Note that the solid line shows the current flowing

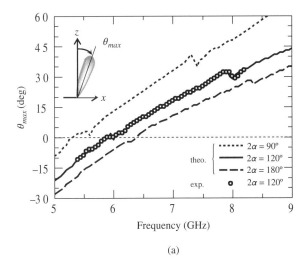

(a)

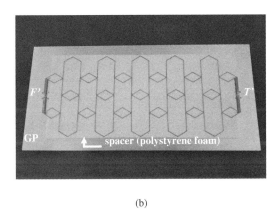

(b)

Figure 14.3 Beam direction and fabricated antenna. (a) Beam direction $\theta = \theta_{max}$ as a function of frequency f. (b) Fabricated antenna. (Reproduced from Ref. [13] with permission from IEEE.)

along the upper paths (AB_1C_1D, EF_1G_1H, IJ_1K_1L, MN_1O_1P, and QR_1S_1T, as illustrated in the inset), while the dotted line shows the current flowing along the lower paths (AB_2C_2D, EF_2G_2H, IJ_2K_2L, MN_2O_2P, and QR_2S_2T).

At the singular gain frequency, the currents along AB_mC_mD ($m = 1, 2$), forming loop 1 (LP-1), are larger than those along the remaining paths (loops). This means that the bent radiation elements of LP-1, that is, B_mC_m ($m = 1, 2$), make a greater contribution to the formation of the radiation pattern, compared with those of the other bent radiation elements. At the nonsingular gain frequency of 7 GHz, the current is distributed almost equally along all the paths, resulting in a nearly equal contribution by all the paths to the formation of the radiation pattern. It follows that the array effect for the radiation pattern at a nonsingular gain frequency is stronger than that at a singular gain frequency. This causes a difference in the radiation

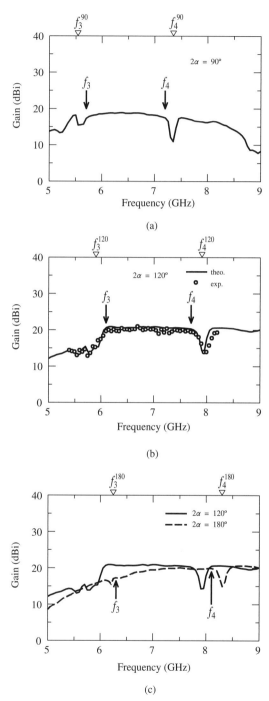

Figure 14.4   Frequency response of the gain in the beam direction θ_{max}. (a) $2\alpha = 90°$. (b) $2\alpha = 120°$. (c) $2\alpha = 180°$, together with $2\alpha = 120°$. (Reproduced from Ref. [13] with permission from IEEE.)

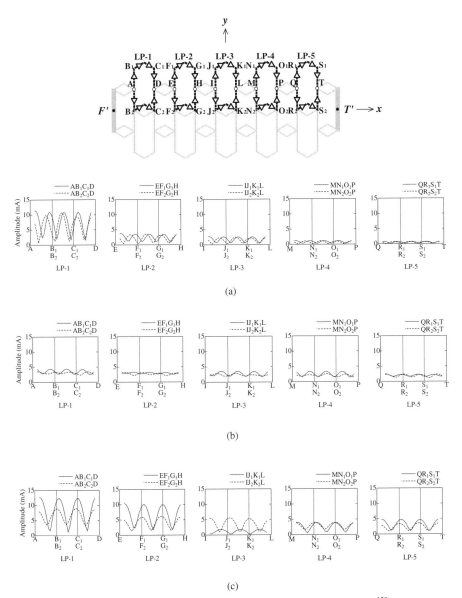

Figure 14.5 Amplitude of the current distribution for bend angle $2\alpha = 120°$. (a) At $f_4^{120} = 7.9$ GHz (singular gain frequency). (b) At 7 GHz (nonsingular gain frequency). (c) At 5.9 GHz (singular gain frequency). (Reproduced from Ref. [13] with permission from IEEE.)

pattern; the HPBW at a singular gain frequency is wider than that at a nonsingular gain frequency. As a result, the gain at f_4^{120} has a dip in the frequency response curve. A similar nonuniform current distribution is observed at $f_3^{120} = 5.9$ GHz, as shown in Fig. 14.5c, where the effect of the $3\lambda_g$-loop resonance appears on the first two loops, LP-1 and LP-2, causing a decrease in the gain.

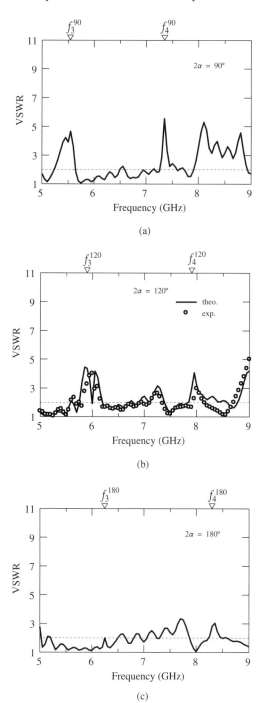

Figure 14.6 VSWR Frequency response. (a) $2\alpha = 90°$. (b) $2\alpha = 120°$. (c) $2\alpha = 180°$. (Reproduced from Ref. [13] with permission from IEEE.)

14.3 VSWR CHARACTERISTIC

Input impedance matching for the R-GAA to a R_0-ohm feed line is performed as follows: (1) the pair of y-directed straight feed lines connected to starting point F' (input line pair) is configured to have a characteristic impedance of $2R_0\,\Omega$ by selecting the wire radius ρ_1 [or equivalent strip width w_1 for a strip grid array], (2) the radiation elements (bent lines), and the feed lines (parallel to the y-axis) after the input line pair are configured to have a characteristic impedance of $4R_0\,\Omega$ by selecting the wire radius ρ_2 [or equivalent strip width w_2], (3) the pair of y-directed straight feed lines connected to the ending point T' (output line pair) is configured to have a characteristic impedance of $2R_0\,\Omega$ by selecting the radius ρ_1 [or equivalent strip width w_1], as with the input line pair, and (4) the output current flowing from the ending point T' to the ground plane is terminated through an $R_0 = 50\,\Omega$ resistor. Note that the radii $\rho_1 = 0.7$ mm ($w_1 = 2.8$ mm) and $\rho_2 = 0.2$ mm ($w_2 = 0.8$ mm) in Table 14.1 are those of commercially available wire, and lead to a low VSWR frequency characteristic for a $50\,\Omega$ transmission line.

Figure 14.6 shows the frequency response of the VSWR for $2\alpha = 90°$, $120°$, and $180°$. It is found that the VSWR between $f_3^{2\alpha}$ and $f_4^{2\alpha}$ is acceptable for practical applications. The high values at and around frequencies $f_3^{2\alpha}$ and $f_4^{2\alpha}$ are attributed to $3\lambda_g$ and $4\lambda_g$ loop resonances.

REFERENCES

1. J. D. Kraus, A backward angle-fire array antenna. *IEEE Trans. Antennas Propag.*, vol. AP-12, no. 1, pp. 48–50, 1964.
2. J. D. Kraus and R. Marhefka, *Antennas*, 3rd edn, Chapter 16, McGraw-Hill, 2002.
3. H. Nakano, I. Oshima, H. Mimaki, J. Yamauchi, and K. Hirose, Numerical analysis of a grid array antenna. *Proc. ICCS'94*, Singapore, vol. 2, 1994, pp. 700–704.
4. R. Harrington, *Time-harmonic electromagnetic fields*, NY: McGraw-Hill, 1961.
5. H. Nakano, T. Kawano, H. Mimaki, J. Yamauchi, and K. Hirose, Grid array antennas. In: *Proc. International Conference on Communication Systems and IEEE International Workshop on Intelligent Signal Processing and Communication Systems*, Singapore, vol. 1, November 1996, pp. 307–311.
6. H. Nakano, T. Kawano, Y. Kozono, and J. Yamauchi, A fast MOM calculation technique using sinusoidal basis and testing functions for a wire on a dielectric substrate and its application to meander loop and grid array antennas. *IEEE Trans. Antennas Propag.*, vol. 53, no. 10, pp. 3300–3307, 2005.
7. H. Nakano, H. Irie, H. Mimaki, and J. Yamauchi, A strip grid array antenna. In: *Proc. APMC, 2002, Kyoto, Japan*, vol. 3, November 2002, pp. 1599–1603.
8. R. Conti, J. Toth, T. Dowling, and J. Weiss, The wire-grid microstrip antenna. *IEEE Trans. Antennas Propag.*, vol. AP-29, no. 1, pp. 157–166, 1981.
9. L. Zhang, W. Zhang, and Y. P. Zhang, Microstrip grid and comb array antennas. *IEEE Trans. Antennas Propag.*, vol. 59, no. 11, pp. 4077–4084, 2011.
10. H. Nakano, H. Osada, and J. Yamauchi, Strip-type grid array antenna with a two-layer rear-space structure. In: *Proc. 7th ISAPE, Guilin, China*, October 2006, pp. 58–61.
11. C. Xing, C. Kain, and H. Kama, A microstrip grid array antenna optimized by a parallel genetic algorithm. *Microw. Opt. Technol. Lett.*, vol. 50, no. 11, pp. 2976–2978, 2008.
12. B. Zhang, L. Zhang, and Y. P. Zhang, Design of low cost linearly-polarized microstrip grid array antenna for 24 GHz doppler sensors. *Proc. 2011 IEEE MWP*, Singapore, October 2011, pp. 93–96.
13. H. Nakano, Y. Iitsuka, and J. Yamauchi, Rhombic grid array antenna. *IEEE Trans. Antennas Propag.*, vol. AP-61, no. 5, pp. 2482–2489, 2013.

Chapter 15

Circularly Polarized Grid Array Antenna

In contrast to the frequency-scanning linearly polarized (LP) beam forming discussed in Chapter 14, frequency-scanning circularly polarized (CP) beam forming is realized in Ref. 1, where spiral elements are adopted as the radiation elements. An issue with this spiral-based CP beam forming is that the gain bandwidth is narrow (7.5%). Another issue is that the CP spiral radiation elements and the feed lines do not lie in the same plane, which complicates the antenna fabrication and assembly. This chapter presents a different CP grid array antenna that resolves these issues, allowing it to be used for a wide range of applications, such as in CP radar and satellite communications systems [2].

15.1 CONFIGURATION OF A PROTOTYPE LOOP-BASED CP GAA$_{EDG}$

Figure 15.1a and b shows, respectively, the perspective and side views of a prototype loop-based edge-excitation grid array antenna (abbreviated as the Pro-GAA$_{EDG}$) that realizes CP-beam frequency scanning. This antenna is constructed from strip lines [with widths w_1 and w_2 ($< w_1$)] and fed from point F'. The antenna elements are located at height H (which is very small relative to the operating wavelength) above the conducting ground plane (GP). The radiation elements (labeled mn) are N open loops, as shown in Fig. 15.1d, connected to the x-directed strips by short-strip lines of length ΔS_{LP}, as shown in Fig. 15.1e.

The y-directed strips are used to feed the loops, whose radii are r_{LP}. At a test frequency of 6 GHz (with a free-space wavelength of λ_6), the spacing L_x is one-half wavelength ($0.5\lambda_6 = 2.5$ cm), and the spacing L_y is one wavelength ($1\lambda_6$). Point F', the starting point of the grid, is located on the left edge, to which the inner conductor of a coaxial feed line is connected (note that the outer conductor of the coaxial

Low-Profile Natural and Metamaterial Antennas: Analysis Methods and Applications, First Edition.
Hisamatsu Nakano.
© 2016 The Institute of Electrical and Electronics Engineers, Inc. Published 2016 by John Wiley & Sons, Inc.

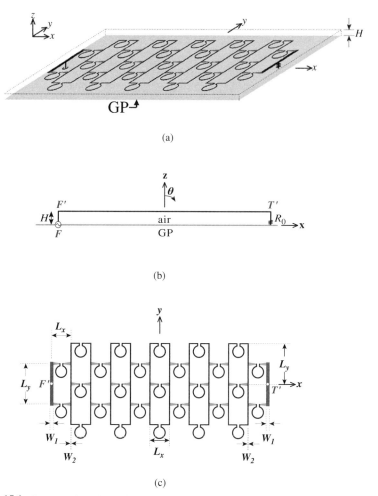

Figure 15.1   Prototype loop-based CP grid array antenna (Pro-GAA$_{EDG}$). (a) Perspective view.
(b) Side view. (c) Top view. (d) Numbering for elements. (e) Detailed view of the open loops.
(Reproduced from Ref. [2] with permission from IEEE.)

feed line is soldered to the ground plane). Point T', the ending point at the right
edge, is shorted to the ground plane (GP) through a resistive load R_0, in order for
the current on the grid array to flow in a traveling wave fashion from point F' to
point T'.

The two pairs of y-directed straight feed strips connected to point F' (input line
pair) and point T' (output line pair) have the same width, w_1. Also, note that the path
length from input point a to output point b across the interval L_x [see Fig. 15.1e],
called the S-length, is denoted as S_{ab}. The parameters that remain fixed throughout
this chapter are listed in Table 15.1. The radius of the open loops r_{LP} is varied subject
to the objectives of the discussion.

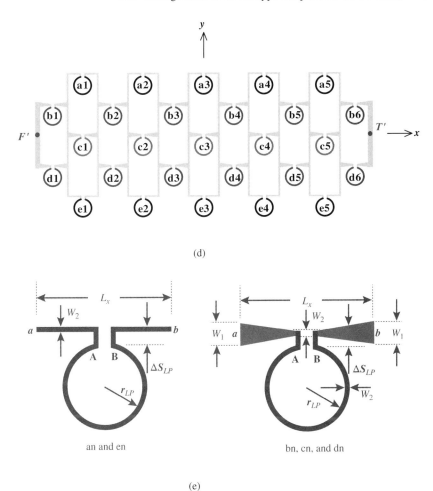

(d)

an and en   bn, cn, and dn

(e)

Figure 15.1   (*Continued*)

Table 15.1   Parameters

Symbol	Value	Unit
N	27	
L_x	25	mm
L_y	50	mm
H	2.5	mm
w_1	2.8	mm
w_2	0.8	mm
ΔS_{LP}	2.5	mm
R_0	50	Ω

Source: Reproduced from Ref. 2 with permission from IEEE.

15.2   RADIATION CHARACTERISTICS OF THE PROTOTYPE LOOP-BASED CP GAA$_{EDG}$

The ground plane, GP, is assumed to be of infinite extent. When the S-length S_{ab} is chosen to be $0.5\lambda_6 \times 3 = 1.5\lambda_6$, in-phase excitation for the loops is realized at a frequency close to 6 GHz, as long as the guided wavelength λ_g is close to the free-space wavelength λ_6 (= 5 cm). Table 15.2 summarizes the S-length S_{ab} for three loop radii r_{LP}, together with the open loop length L_{AB} from point A to point B; $r_{LP} = 9.5$ mm $\equiv r_{LP0}$ (called the reference loop radius), 8.5 mm $\equiv r_{LP-} < r_{LP0}$, and 10.5 mm $\equiv r_{LP+} > r_{LP0}$.

As the frequency is increased from 6 GHz, the electrical S-length S_{ab}/λ (λ is the free-space wavelength at the operating frequency) becomes longer because of the decrease in wavelength λ. This affects the initial in-phase excitation at the input points of neighboring loops; the excitation phases become regressive (delayed) toward the ending point T'. As a result, the radiation becomes tilted toward the positive x-space ($+\theta$ direction), that is, the antenna forms a forward beam. Note that, as the frequency is decreased from 6 GHz, the electrical S-length S_{ab}/λ

Table 15.2   S-Length S_{ab} and Open Loop Length L_{AB}

r_{LP}	S_{ab}	L_{AB}
$r_{LP-} = 8.5$ mm	74.6 mm	49.0 mm
	$(1.49\lambda_6)$	$(0.98\lambda_6)$
$r_{LP0} = 9.5$ mm	79.8 mm	54.7 mm
	$(1.60\lambda_6)$	$(1.10\lambda_6)$
$r_{LP+} = 10.5$ mm	85.0 mm	60.5 mm
	$(1.70\lambda_6)$	$(1.21\lambda_6)$

Source: Reproduced from Ref. 2 with permission from IEEE.

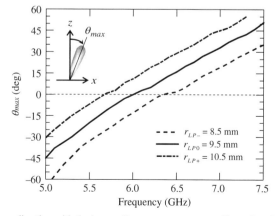

Figure 15.2   Beam direction with the loop radius r_{LP} as a parameter. (Reproduced from Ref. [2] with permission from IEEE.)

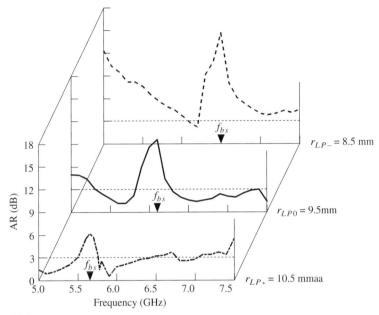

Figure 15.3 Frequency response of the axial ratio, with the loop radius r_{LP} as a parameter. (Reproduced from Ref. [2] with permission from IEEE.)

becomes shorter, that is, the phases at the input points of neighboring loops become progressive toward the ending point T'. Thus, the antenna generates a tilted beam toward the negative x-space ($-\theta$ direction); in other words, the antenna forms a backward beam.

Figure 15.2 shows the frequency response of the beam tilt angle (scanning-beam angle) with the loop radius r_{LP} as a parameter. The frequency response curve for the larger radius r_{LP+} is shifted to the left relative to the response for the reference radius r_{LP0}. This is due to the fact that the S-length S_{ab} from point a to point b for r_{LP+} is physically longer than the S-length for r_{LP0}. Note that the leftward shift in the frequency response curve can be used to reduce the antenna size, that is, the antenna can be made smaller by applying a scaling technique to this frequency response curve. Conversely, the frequency response curve for the smaller loop radius r_{LP-} is shifted to the right relative to the response for the reference loop radius r_{LP0}, because the S-length S_{ab} for r_{LP-} is physically shorter than the S-length for the reference loop radius r_{LP0}.

The axial ratio (AR) in the beam direction with the loop radius r_{LP} as a parameter is shown in Figure 15.3, where f_{bs} denotes the frequency at which broadside radiation occurs. It is found that the AR for r_{LP0} is undesirable at frequencies around f_{bs}, showing values of greater than 3 dB. This behavior is closely related to the current distribution.

The currents in Fig. 15.4a and c flow in a traveling wave or a quasi-traveling wave fashion, resulting in a good axial ratio. On the other hand, the current in Fig. 15.4b shows that the loop resonates. This is attributed to the fact that the open

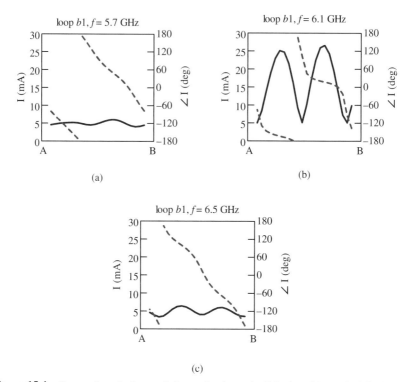

Figure 15.4 Current along the loop path from point A to point B for loop $b1$ near the left edge, where the loop radius $r_{LP} = r_{LP0} = 9.5$ mm. (a) At a low nonbroadside-radiation frequency of 5.7 GHz. (b) At the broadside-radiation frequency of 6.1 GHz. (c) At a high nonbroadside-radiation frequency of 6.5 GHz. Solid and dashed lines illustrate amplitude and phase, respectively. (Reproduced from Ref. [2] with permission from IEEE.)

loop length L_{AB} is approximately one wavelength at 6.1 GHz. The standing wave current along the loop indicates that an undesirable backward current flowing along the loop from point B to point A exists, producing a cross-polarized radiation field [undesirable left-handed (LH) CP wave], leading to a large axial ratio.

The results in Figs. 15.3 and 15.4, together with the gain characteristic shown in Fig. 15.5 [where the right-handed (RH) CP gain, G_R, decreases at $f_{bs} = 6.1$ GHz], indicate that the Pro-GAA$_{\text{EDG}}$ can be applied to systems where a CP beam in the direction normal to the antenna plane is *not* required. For example, the Pro-GAA$_{\text{EDG}}$ can be applied to a surveillance radar system or a collision avoidance radar system: a Pro-GAA$_{\text{EDG}}$ could be installed at the rear of a vehicle (such as a ship or car) to collect information about other vehicles that are behind the vehicle (at an oblique angle). Note that the x–y plane in Fig. 15.1a would be perpendicular

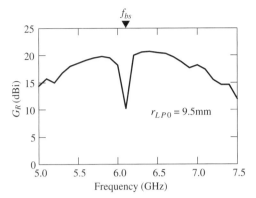

Figure 15.5 Frequency response of the gain for the right-handed CP wave, G_R, where the open loop radius is $r_{LP} = r_{LP0} = 9.5$ mm. (Reproduced from Ref. [2] with permission from IEEE.)

to the surface of the water when used on a ship, and perpendicular to the surface of the road when used on a land vehicle.

15.3 CONFIGURATION OF AN ADVANCED LOOP-BASED CP GAA$_{EDG}$

The Pro-GAA$_{EDG}$ has a nontraveling-wave current across a small frequency region around f_{bs}, resulting in the broadside radiation being elliptically or linelly polarized. This frequency region is called the non-CP region. The open loop length L_{AB} is approximately one wavelength across the non-CP region, and hence the loops resonate in this frequency range. To avoid the resonance, small filaments (perturbation elements used as reactive elements) are added to the loops [2], thereby transforming the nontraveling-wave current observed across the non-CP region into a traveling-wave or quasi-traveling-wave current. This advanced loop-based CP GAA$_{EDG}$ with perturbation elements is abbreviated as the Adv-GAA$_{EDG}$.

A design example for the Adv-GAA$_{EDG}$ is explained using two steps, where the reference loop radius ($r_{LP} = r_{LP0} = 9.5$ mm) is used, together with the parameters shown in Table 15.1. In the first step, the rows (a, b, . . . , e) of the loops that need the perturbation elements are determined on the basis of the current distribution analysis for the Pro-GAA$_{EDG}$ at $f_{bs} = 6.1$ GHz. The analysis reveals that the loops near the starting point F' in rows b, c, and d exhibit significant nontraveling-wave current behavior [2]. This fact infers that the nontraveling wave currents will be reduced when perturbation elements (of length Δp) are added to the loops in rows b, c, and d, as shown in Fig. 15.6.

In the second step, the perturbation element length Δp is optimized to obtain a traveling wave current at $f_{bs} = 6.1$ GHz. Figure 15.7 shows the current distribution for $\Delta p = 3$ mm. As seen from the phase distribution along the open loop path from point A to point B for each loop, the current flows in an almost traveling-wave fashion, as desired. A comparison of the current distribution for the Adv-GAA$_{EDG}$

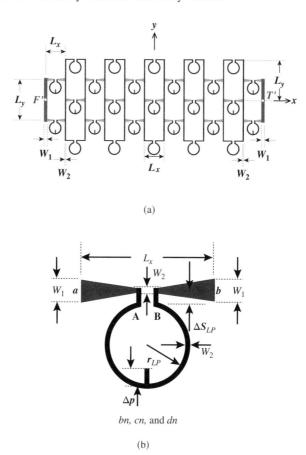

(a)

bn, cn, and dn

(b)

Figure 15.6 Advanced loop-based CP GAA$_{EDG}$ (abbreviated as Adv-GAA$_{EDG}$), where the open loop radius is $r_{LP} = r_{LP0} = 9.5$ mm. (a) Top view. (b) Loops bn, cn, and dn, where n is the column number shown in Fig. 15.1d. (Reproduced from Ref. [2] with permission from IEEE.)

(Fig. 15.7) with that for the Pro-GAA$_{EDG}$ illustrated in Ref. [2] clearly shows that the perturbation elements are effective for transforming the standing wave current observed in the Pro-GAA$_{EDG}$ into a traveling wave current.

15.4 RADIATION CHARACTERISTICS OF THE ADVANCED LOOP-BASED CP GAA$_{EDG}$

The radiation pattern for the Adv-GAA$_{EDG}$ (Fig. 15.8a) is compared with that for the Pro-GAA$_{EDG}$ (Fig. 15.8b), where E_R and E_L denote the RH CP radiation field component and the LH CP radiation field component, respectively. It is found that a large difference exists between the broadside radiation patterns at 6.1 GHz. By achieving CP radiation in the broadside direction, the Adv-GAA$_{EDG}$ realizes seamless RH CP radiation. Experimental results using a fabricated antenna [2], shown in

—— I : amplitude

--- ∠I : phase

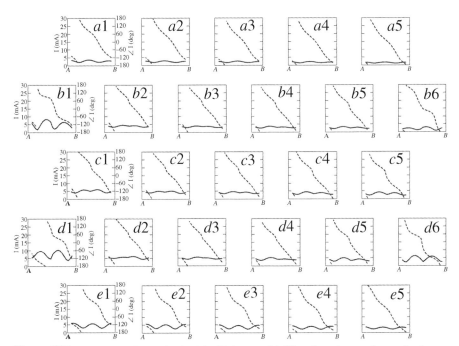

Figure 15.7 Current distribution for the Adv-GAA$_{EDG}$ at 6.1 GHz, where the open loop radius is $r_{LP} = r_{LP0} = 9.5$ mm and $\Delta p = 3$ mm. (Reproduced from Ref. [2] with permission from IEEE.)

Fig. 15.9, confirm this fact, where the infinite ground plane for the analysis is approximated using a finite conducting plane of $3.5\lambda_6 \times 6.5\lambda_6$, and a polystyrene foam layer (relative permittivity $\varepsilon_r \approx 1$) is inserted into the space between the grid array plane and the ground plane, 2.5 mm, to support the grid array.

The frequency response of the axial ratio (AR) for the Adv-GAA$_{EDG}$ in the beam direction is shown in Fig. 15.10. The axial ratio is found to be less than 3 dB over a frequency range of 5.3–7.3 GHz (32%), where the beam direction varies from $\theta_{max} = -28°$ to +48°, as shown in Fig. 15.11. Figure 15.12 shows the input performance for the Adv-GAA$_{EDG}$ in terms of the VSWR relative to 50 Ω. The Adv-GAA$_{EDG}$ has a low VSWR across the aforementioned axial ratio bandwidth (from 5.3 GHz to 7.3 GHz), as desired. Note that the strip widths $w1$ and w_2 shown in Table 15.1 are chosen such that the input impedance matches a 50 Ω feed line.

The fact that the broadside radiation for the Adv-GAA$_{EDG}$ is circularly polarized leads to a seamless gain characteristic. Figure 15.13 shows the gain for an RH CP wave, G_R, where the loss due to impedance mismatch is not included. The

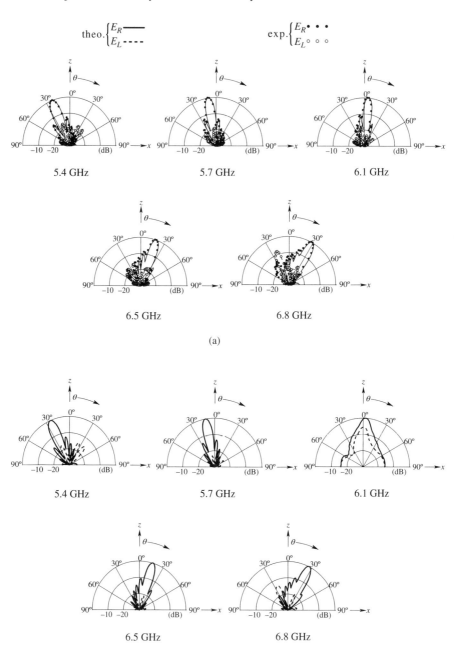

Figure 15.8 Radiation patterns. (a) Adv-GAA$_{EDG}$. (b) Pro-GAA$_{EDG}$. (Reproduced from Ref. [2] with permission from IEEE.)

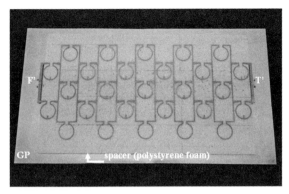

Figure 15.9 Fabricated Adv-GAA_EDG. (Reproduced from Ref. [2] with permission from IEEE.)

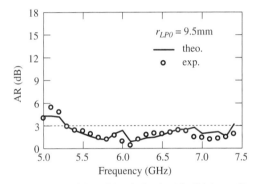

Figure 15.10 Frequency response of the axial ratio for the Adv-GAA_EDG. (Reproduced from Ref. [2] with permission from IEEE.)

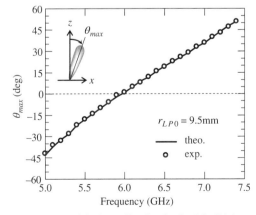

Figure 15.11 Frequency response of the beam direction for the Adv-GAA_EDG. (Reproduced from Ref. [2] with permission from IEEE.)

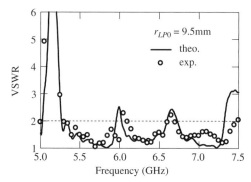

Figure 15.12   Frequency response of the VSWR for the Adv-GAA$_{EDG}$. The VSWR value is relative to 50 Ω. (Reproduced from Ref. [2] with permission from IEEE.)

Adv-GAA$_{EDG}$ has a G_R bandwidth of 23% (from 5.4 GHz to 6.8 GHz for a 3 dB reduced gain criterion), which is approximately three times wider than that for a spiral-based CP GAA$_{EDG}$ [1]. Note that the radiation efficiency becomes larger as the frequency increases, showing a value of more than 50% across the gain bandwidth (from 5.4 GHz to 6.8 GHz) [2].

Exercise

Design a high directivity (of more than 20 dB at 12 GHz) CP antenna using a grid array structure.

Answer   Figure 15.14 shows a grid array antenna composed of four grid subarrays (made of wire of radius a) [3,4]. Points F_1, F_2, F_3, and F_4 are the feed points, which are fed with a 90°

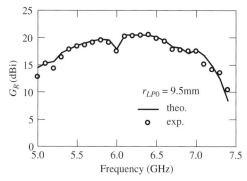

Figure 15.13   Frequency response of the gain G_R for the Adv-GAA$_{EDG}$. (Reproduced from Ref. [2] with permission from IEEE.)

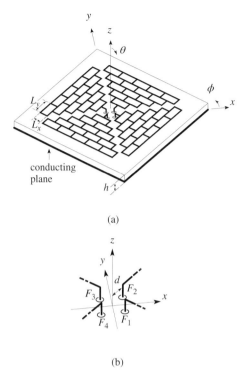

(a)

(b)

Figure 15.14   Array antenna composed of four grid subarrays. (Reproduced from Ref. [3] with permission from IEEE.)

phase difference between neighboring feed points [5]. The parameters used are shown in Table 15.3, where λ_{12} is the free-space wavelength at 12 GHz. Figure 15.15 shows the frequency response of the directivity, which meets the design requirement. The radiation patterns are also shown in Fig. 15.16. ∎

Table 15.3   Configuration Parameters

Symbol	Value
h	$0.07\lambda_{12}$
L_x	$0.53\lambda_{12}$
L_y	$1.06\lambda_{12}$
$2a$	$0.012\lambda_{12}$
d	$0.07\lambda_{12}$

Source: Reproduced from Ref. 3 with permission from IEEE.

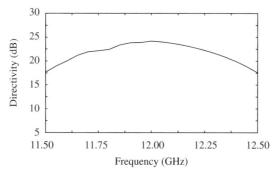

Figure 15.15   Frequency response of the directivity. (Reproduced from Ref. [3] with permission from IEEE.)

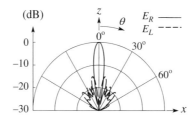

Figure 15.16   Radiation pattern. (Reproduced from Ref. [3] with permission from IEEE.)

REFERENCES

1. Y. Iitsuka, J. Yamauchi, and H. Nakano, Circularly polarized spiral-grid array antenna for beam scanning. In: *Proc. Int. Symp. Antennas and Propag., (ISAP)*, Bangkok, October 2009, pp. 5–8.
2. H. Nakano, Y. Iitsuka, and J. Yamauchi, Loop-based circularly polarized grid array antenna with edge excitation. *IEEE Trans. Antennas Propag.*, vol. AP-61, no. 8, pp. 4045–4053, 2013.
3. T. Kawano and H. Nakano, Numerical analysis of a bowtie-shaped grid array antenna. In: *Proc. IEEE APWC 15*, Torino, September 2015, pp. 250–251.
4. T. Kawano and H. Nakano, A bowtie-shaped grid array antenna radiating linearly and circularly polarized beams. In: *Proc. IEEE APS Int. Sympo. Antennas Propag.*, Vancouver, July 2015, pp. 1904–1905.
5. H. Nakano, N. Suzuki, and J. Yamauchi, Mesh antenna (IV)—Characteristics for circular polarization. In: *Proc. IEICE Society Conference*, 1998, p. B-1-128.

Earth–Satellite and Satellite–Satellite Communications Antennas

Chapter 16

Monofilar Spiral Antenna Array

The spirals in Refs 1 and 2 are composed of two arms and the spiral in Ref. 3 is composed of multiple arms. The feed systems for these spirals are complicated because a balun circuit or an excitation circuit consisting of power dividers and phase shifters is required. In contrast, the spiral element to be discussed here is composed of a single arm (called a monofilar spiral antenna) and does not require a balun circuit. Consequently, the monofilar spiral has a simpler feed system than the two- and multiarm spirals.

In this chapter, first, the radiation characteristics of the monofilar spiral [4] are investigated as a circularly polarized (CP) tilted-beam element, and, second, the performance of an array antenna composed of four CP tilted-beam monofilar spirals is discussed for application to satellite communications.

16.1 TILTED-BEAM MONOFILAR SPIRAL ANTENNA

Figure 16.1 shows a monofilar spiral antenna made of a wire of diameter $2a$. The curved arm section at height h above a conducting plane (ground plane, GP) is defined by the Archimedean spiral function, that is, the radial distance r' from the center O' to a point on the arm is given by $a_{sp}\phi_w$, where a_{sp} is the spiral constant and ϕ_w is the winding angle starting at ϕ_{st} and ending at ϕ_{end}. The antenna is fed from a coaxial line.

A tilted beam is formed by superimposing radiation field RF_1 on radiation field RF_2, where RF_1 is the radiation field from the first active region of approximately one-wavelength circumference on the spiral plane and RF_2 is the radiation field from the second active region of approximately two-wavelength circumference. RF_1 has approximately in-phase fields at two symmetrical points with respect to the z-axis, whereas RF_2 has a phase difference at these points. The superimposition of

Low-Profile Natural and Metamaterial Antennas: Analysis Methods and Applications, First Edition. Hisamatsu Nakano.

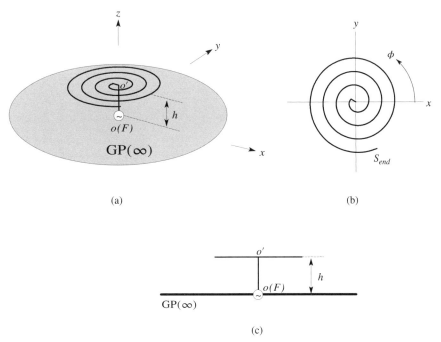

(a)

(b)

(c)

Figure 16.1   Monofilar spiral antenna. (a) Perspective view. (b) Top view. (c) Side view.

RF_2 onto RF_1 leads to a tilted beam. On the basis of this mechanism, the outermost periphery (circumference) of the curved arm section must be more than two wavelengths.

The parameters fixed throughout this chapter are as follows: wire diameter $2a = 0.012\lambda_6$, antenna height $h = 0.25\lambda_6$, spiral constant $a_{sp} = 0.0153\lambda_6$/rad, starting angle $\phi_{st} = 2.6$ rad, and ending angle $\phi_{end} = 24.0$ rad, where λ_6 is the wavelength at a test (design) frequency of $f_0 = 6$ GHz. The circumference of the spiral, defined as $C = 2\pi a_{sp}\phi_{end}$, is $2.3\lambda_6$, corresponding to a diameter of $0.73\lambda_6$. This diameter is less than $1\lambda_6$ and contributes to suppressing grating lobes in the spiral array antenna discussed in the next subsection. Note that the ground plane is assumed to be of infinite extent.

Figure 16.2a shows the current distribution $I = I_r + jI_i$ at frequency $f_0 = 6$ GHz, which is obtained using the method of moments in Section 2. The arm length from feed point O to the arm ending point, S_{end}, is approximately $4.64\lambda_6$. It is found that a decaying traveling wave current flows toward the arm end. This means that the current reflected from the arm end is small, which implies that the input impedance remains unchanged when the frequency changes. It is also found in Fig. 16.2b that the phase progression is close to that in free space.

The radiation at frequency $f_0 = 6$ GHz becomes maximal off the z-axis: $(\theta, \phi) = (28°, 232°) \equiv (\theta_{max}, \phi_{max})$ at $f_0 = 6$ GHz. Figure 16.3 shows the radiation

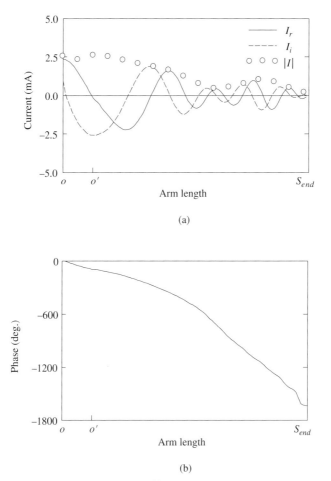

Figure 16.2 Current distribution. (a) I_r, I_i, and $|I|$. (b) Phase progression. (Reproduced from Ref. [4] with permission from IEEE.)

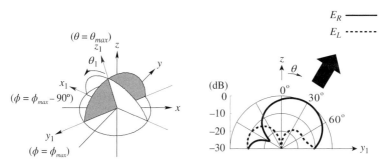

Figure 16.3 Radiation pattern as a function of angle θ, where ϕ is fixed at $232° \equiv \phi_{max}$. (Reproduced from Ref. [4] with permission from IEEE.)

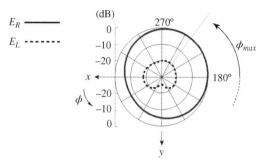

Figure 16.4 Radiation pattern as a function of azimuth angle ϕ, where θ is fixed at $28° \equiv \theta_{max}$. (Reproduced from Ref. [4] with permission from IEEE.)

pattern as a function of θ, where ϕ is fixed at ϕ_{max} ($\theta - \phi_{max}$ plane) and Fig. 16.4 shows the radiation pattern as a function of ϕ, where θ is fixed at θ_{max} ($\theta_{max} - \phi$ plane). The half-power beam width (HPBW) of the copolar component E_R in these patterns is found to be wide, as desired for an array antenna element. Note that E_R and E_L in these figures show a right-handed CP wave component and a left-handed CP wave component, respectively.

Figure 16.5 shows the frequency response of the gain in a fixed direction of $(\theta, \phi) = (\theta_{max}, \phi_{max}) = (28°, 232°)$, that is, the beam direction observed at f_0 GHz. A gain of 8.2 dBi is obtained at f_0, which is comparable to the gain of a two-wire spiral antenna backed by a conducting plane reflector [5]. The frequency bandwidth for a 1 dB gain drop from the maximum value is approximately 12%. The gain drop is mainly due to the deviation of the beam from the fixed direction $(\theta_{max}, \phi_{max})$ with change in frequency (see Fig. 16.6).

The axial ratio in a beam direction of $(\theta, \phi) = (\theta_{max}, \phi_{max}) = (28°, 232°)$ is shown in Fig. 16.7. The frequency bandwidth for a 3 dB axial ratio criterion is

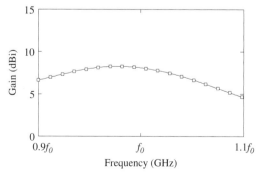

Figure 16.5 Frequency response of the gain in a fixed direction of $(\theta, \phi) = (\theta_{max}, \phi_{max}) = (28°, 232°)$. (Reproduced from Ref. [4] with permission from IEEE.)

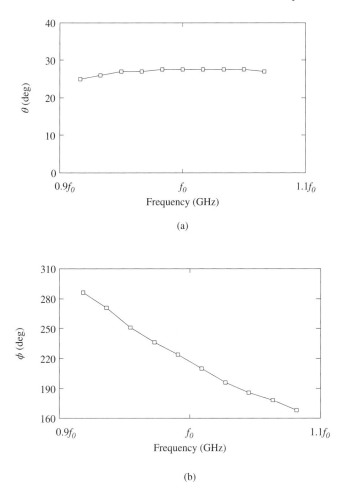

Figure 16.6 Frequency response of the beam direction. (a) Variation in θ. (b) Variation in ϕ. (Reproduced from Ref. [4] with permission from IEEE.)

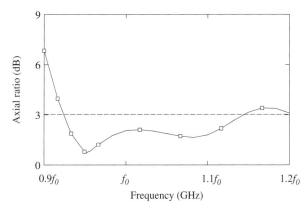

Figure 16.7 Frequency response of the axial ratio in a beam direction of $(\theta, \phi) = (\theta_{max}, \phi_{max}) = (28°, 232°)$. (Reproduced from Ref. [4] with permission from IEEE.)

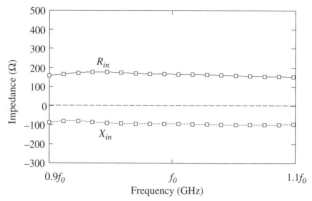

Figure 16.8   Frequency response of the input impedance. (Reproduced from Ref. [4] with permission from IEEE.)

approximately 23%. This bandwidth is slightly wider than that for the two-wire spiral backed by a conducting plane reflector (16% for a $1.4\lambda_0$ circumference [5]). Figure 16.8 shows the frequency response of the input impedance $Z_{in} = R_{in} + jX_{in}$. The resistance value of the input impedance is approximately $150\,\Omega$. The variation in the input impedance with change in frequency is small across a wide frequency range, because a decaying traveling wave current is maintained along the arm.

16.2   TILTED CP FAN BEAM

This section discusses the formation of a tilted CP *fan beam* using an array of monofilar spiral elements, each element having the antenna characteristics revealed in Section 16.1. This beam is particularly useful for satellite communications applications, because polarization alignment between the transmitting and receiving antennas is not necessary. Note that a tilted LP fan beam is discussed in Chapter 12.

Figure 16.9 shows an array composed of four monofilar spirals, where the spacing between neighboring elements is chosen to be less than $1\lambda_6$ ($d = 0.8\lambda_6$). To

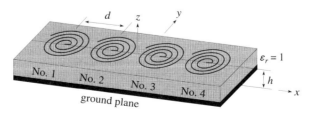

Figure 16.9   Array composed of four monofilar spirals.

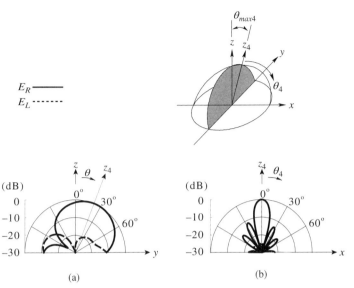

Figure 16.10 Radiation from a four-spiral array at frequency $f_0 = 6$ GHz. (a) In the y–z plane. (b) In the x–z_4 plane. (Reproduced from Ref. [4] with permission from IEEE.)

form a CP fan beam in the y–z plane ($\phi = 90°$), the four spiral element are rotated around their axes by $360° - \phi_{max} + 90°$ ($= 218°$) in the x–y plane and excited with the same amplitude and the same phase.

The radiation from the array at frequency f_0 forms a tilted CP fan beam in the y–z plane with a tilt angle of $24° \equiv \theta_{max4}$ and a HPBW of approximately $70°$, as shown in Fig. 16.10a. This is almost the same radiation pattern as that shown in Fig. 16.3. On the other hand, a narrow HPBW is realized for the radiation pattern in the x–z_4 plane, due to array effects. Note that the z_4-axis is the axis specified by angle θ_{max4}, as shown in the inset of Fig. 16.10.

Figure 16.11 shows the beam direction (θ, ϕ) as a function of frequency. Figures 16.12 and 6.13 depict the frequency response of the axial ratio and gain, respectively, for a fixed beam direction of $(\theta, \phi) = (\theta_{max4}, 90°)$. Recall that θ_{max4} is $24°$ observed at $f_0 = 6$ GHz. It is found that the gain is approximately 6 dB higher at f_0 compared to the gain of a single spiral element. It is also found that the axial ratio bandwidth for the array (approximately 23% bandwidth) is the same as that for a single element.

Figure 16.14 depicts the frequency response of the input impedance for each spiral element of the array, which shows a behavior similar to that observed for the single monofilar spiral antenna (see Fig. 16.8). The variation in the input impedance with frequency is found to be small.

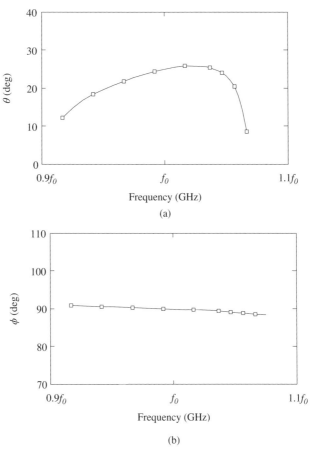

Figure 16.11 Frequency response of the beam direction (θ, ϕ) for the four-spiral array. (a) Direction θ. (b) Direction ϕ. (Reproduced from Ref. [4] with permission from IEEE.)

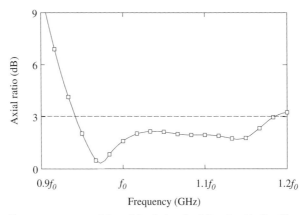

Figure 16.12 Frequency response of the axial ratio in a *fixed* direction $(\theta, \phi) = (\theta_{max4}, 90°)$ for the four-spiral array, where $\theta_{max4} = 24°$. (Reproduced from Ref. [4] with permission from IEEE.)

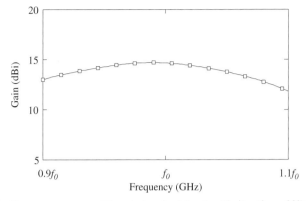

Figure 16.13 Frequency response of the gain in a *fixed* direction $(\theta, \phi) = (\theta_{max4}, 90°)$ for the four-spiral array, where $\theta_{max4} = 24°$. (Reproduced from Ref. [4] with permission from IEEE.)

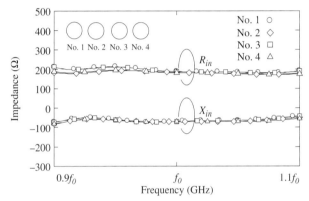

Figure 16.14 Frequency response of the input impedance of each spiral element of the array. (Reproduced from Ref. [4] with permission from IEEE.)

REFERENCES

1. J. A. Kaiser, The Archimedean two-wire spiral antenna. *IRE Trans. Antennas Propag.*, vol. AP-8, no. 3, pp. 312–323, 1960.
2. H. Nakano, H. Yasui, and J. Yamauchi, Numerical analysis of two-arm spiral antennas printed on a finite-size dielectric substrate. *IEEE Trans. Antennas Propag.*, vol. 50, no. 3, pp. 362–370, 2002.
3. L. Shafai, Design of multi-arm multi-mode spiral antennas for directional beams using equivalent array concept. *Electromagnetics*, vol. 14, no. 3/4, pp. 285–304, 1994.
4. H. Nakano, Y. Shinma, and J. Yamauchi, A Monofilar spiral antenna and its array above a ground plane-formation of a circularly polarized tilted fan beam. *IEEE Trans. Antennas Propag.*, vol. 45, no. 10, pp. 1506–1511, 1997.
5. H. Nakano, K. Nogami, S. Arai, H. Mimaki, and J. Yamauchi, A spiral antenna backed by a conducting plane reflector. *IEEE Trans. Antennas Propag.*, vol. AP-34, pp. 791–796, 1986.

Chapter 17

Low-Profile Helical Antenna Array

In this chapter, we design a low-profile high-gain helical antenna array [1] for satellite communications, using the MoM and rotational vector method [2]. Each helical antenna acts as an array element, having a vertical probe, which protrudes into a thin circular cavity (radial transmission line [3,4]). The vertical probe is excited by an electromagnetic (EM) wave traveling in the circular cavity, that is, the helical element absorbs the EM wave power inside the cavity through the vertical probe and radiates a circularly polarized (CP) wave into free space [5]. The antenna gain increases as the number of helical elements is increased. Controlling the excitation amplitude and phase of each helical element is also discussed in this chapter.

17.1 ARRAY ELEMENT

The helical antenna shown in Fig. 17.1a and b is specified by the following parameters: diameter of the conducting circular ground plane D_{GP}, diameter of the helical arm $2r_{helix}$, pitch angle α, number of helical turns N, wire diameter of the helical arm $2a$, starting height of the helical arm above the ground plane h, and antenna height H.

Figure 17.1c shows the current distribution for a long helical antenna ($N = 20$) from the feed point F to the helical arm end S_e, where a frequency of 12 GHz (the wavelength is $25\,\mathrm{mm} \equiv \lambda_{12}$) is used, together with the parameters listed in Table 17.1. It is found that there are two distinct regions: a decaying current region from the feed point to the first minimum point M (C-region) and the remaining current region that is characterized by a relatively constant amplitude (S-region).

In contrast to Fig. 17.1, Fig. 17.2 shows the current distribution of a short helical antenna ($N = 2$) at 12 GHz, where $H = 4.7\,\mathrm{mm} = 0.188\lambda_{12}$. The current along the helical arm is characterized by a single decaying current (C-region). Figure 17.3 shows the radiation patterns at 12 GHz, where E_R and E_L denote the right-handed and left-handed CP wave components, respectively. It is revealed that the radiation

Low-Profile Natural and Metamaterial Antennas: Analysis Methods and Applications, First Edition.
Hisamatsu Nakano.
© 2016 The Institute of Electrical and Electronics Engineers, Inc. Published 2016 by John Wiley & Sons, Inc.

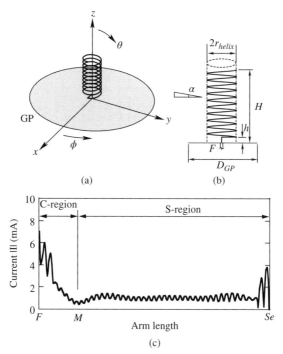

(a)   (b)

(c)

Figure 17.1   Axial mode helical antenna. (a) Perspective view. (b) Side view. (c) Current distribution (amplitude) for $N = 20$ at 12 GHz, where the parameters in Table 17.1 are used.

Table 17.1   Configuration Parameters

Symbol	Value
r_{helix}	3.98 mm
α	4°
$2a$	1.0 mm
h	1.25 mm
D_{GP}	∞

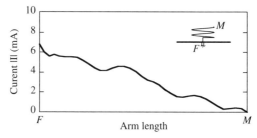

Figure 17.2   Current distribution for a low-profile helical antenna ($N = 2$). (Reproduced from Ref. [1] with permission from IEEE.)

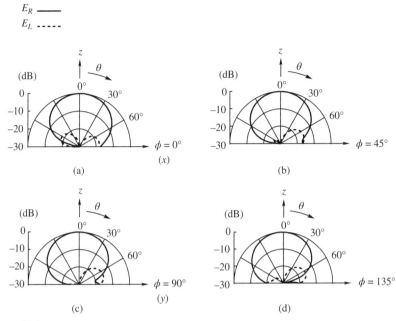

E_R ⎯⎯⎯
E_L ⎯ ⎯ ⎯

Figure 17.3 Radiation patterns for the low-profile helical antenna ($N=2$), where the parameters in Table 17.1 are used. (a) In the $\phi=0°$ plane (x–z plane). (b) In the $\phi=45°$ plane. (c) In the $\phi=90°$ plane (y–z plane). (d) In the $\phi=135°$ plane. (Reproduced from Ref. [1] with permission from IEEE.)

in the positive z-space is circularly polarized and the half-power beamwidth (HPBW) is wide, as desired for an array element. This CP low-profile helical antenna is used as an element for an array antenna in the following subsections.

17.2 ARRAY ANTENNA

17.2.1 Configuration

Figure 17.4 shows a circular cavity (radial transmission line) on which the CP low-profile helical antennas ($N=2$) discussed in Section 17.1 are arrayed. The number of helical elements is $6n$ on the nth circumference (ring), where the spacing between array elements in the circumferential direction is given by $S_{cir}=2\pi$ $(nS_{rad})/(6n)=\pi S_{rad}/3$. Note that nS_{rad} is the radius of the nth circumference.

The distance between the top and bottom plates of the circular cavity, S_W, is chosen to be small with respect to the operating wavelength so that a local electromagnetic (EM) wave traveling along the radial direction from the center of the cavity to the edge (side wall) of the cavity is approximated by a TEM wave. A value of $S_W=7.5$ mm $=0.3\lambda_{12}$ is used throughout this chapter.

The vertical probe of each helical element protrudes into the circular cavity through a small hole in the top plate of the cavity and absorbs the EM wave

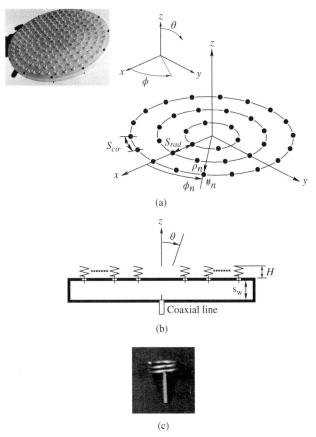

(a)

(b)

(c)

Figure 17.4   Array configuration. (a) Perspective view. (Reproduced from Ref. [1] with permission from IEEE.) (b) Side view. (c) Array element.

traveling inside the cavity. Note that the EM wave is generated from a probe (connected to the inner conductor of a coaxial feed line) located at the center of the bottom plate of the cavity.

17.2.2 Directivity

A high-gain beam in the direction normal to the cavity surface (normal beam) is used for satellite communications. The gain in the normal direction (z-direction) becomes maximal when all the helical elements are excited in-phase. For this, the excitation phase for each helical element is adjusted by mechanically rotating the helical element around its axis (which coincides with the vertical probe) [1]. This phase adjustment technique is designated as the *axis rotation technique* (ART) [6], which is simple and differs from techniques that use phase shifters or phase circuits.

A high-gain beam in a direction off the z-axis is also used for satellite communication, as shown in Fig. 17.5. There is no need to tilt the antenna because the

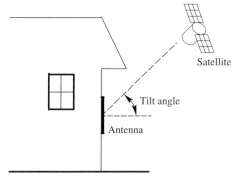

Figure 17.5 Satellite communication using a tilted beam.

beam is already tilted. The tilted beam is realized by gradating the excitation phase of each helical element. Generally, as the beam tilt angle is increased, side lobes become larger, resulting in a decrease in the antenna gain. Therefore, suppressing side lobes is important for avoiding a remarkable reduction in the gain. For this, the spacing between helical elements must be appropriately chosen.

The directivity for the helical antenna array in the beam direction (θ_0, ϕ_0) is calculated using

$$D(\theta_0, \phi_0) = \frac{4\pi|F(\theta_0, \phi_0)|^2}{\int_0^{2\pi} d\phi \int_0^{\pi} |F(\theta, \phi)|^2 \sin\theta \, d\theta} \tag{17.1}$$

where $F(\theta, \phi)$ is the electric field pattern for the helical antenna array and given by

$$F(\theta, \phi) = E_0(\theta, \phi) \sum_{n=1}^{N_a} I_n e^{j[k\rho_n \sin\theta \cos(\phi-\phi_n)+\delta_n]} \tag{17.2}$$

where $E_0(\theta,\phi)$ is the electric field pattern of the helical element, N_a is the total number of helical elements, k is the wavenumber (the phase constant in free space), and ρ_n is the distance from the coordinate origin to helical element n. I_n and δ_n are the excitation amplitude and phase for helical element n, respectively. The phase δ_n is set to be

$$\delta_n = -k\rho_n \sin\theta_0 \cos(\phi_0 - \phi_n) \tag{17.3}$$

Equation (17.1) is calculated using a numerical integration technique, where $E_0(\theta,\phi)$ is approximated by the principal field component E_R in Fig. 17.3, because the cross-polarized component E_L is small. Figure 17.6 shows the calculated directivity as a function of tilt angle from the z-axis, where the following assumptions are used: (1) the helical elements located on a ring of the same circumference have the same excitation amplitude, and (2) the field pattern of the helical element is $E_0 = E_R = (\cos\theta)^{1.7}$. Note that the directivity shown in Fig. 17.6 is obtained under the condition that the diameter of the helical antenna array is held constant (38.5 cm).

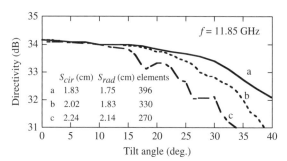

Figure 17.6   Directivity as a function of tilt angle from the z-axis. (Reproduced from Ref. [1] with permission from IEEE.)

17.2.3 Probe Length

The amplitude of the excitation for each helical element is controlled by the length of the vertical probe that protrudes into the cavity. The helical elements on the pth ring (whose circumference is $2\pi p S_{rad}$) have the same protrusion length, which gradually increases, as the ring number p is increased. It is desired that the power of the forward traveling wave from the center of the cavity toward to the edge (side-wall of the cavity) is completely absorbed by the helical elements through the vertical probes and no power exists at the edge of the cavity. For this, the probes of the helical elements on the outermost ring are located at a distance of one-quarter wavelength away from the edge of the cavity at the design frequency.

A method (vector rotation method [2]) presented in Exercise is recommended for obtaining the amplitude and phase required for the helical array element. Note that impedance matching of the coaxial feed line to the cavity is realized by appropriately choosing the length of the probe located at the center of the bottom plate of the cavity.

17.2.4 Radiation Characteristics

We investigate an array antenna composed of 396 helical elements (arranged in 11 rings) with an element spacing of $S_{rad} = 1.75$ cm. Figure 17.7a and b shows the calculated normal beam radiation and the 30°-tilted beam radiation at 11.85 GHz, respectively. The half-power beam width (HPBW) is 3.7° for the normal beam and 4.2° for the tilted beam. The level of the first side lobe is approximately −18 dB for the normal beam and −17 dB for the tilted beam. The axial ratio for both beams is less than 1 dB. The experimental results presented in Ref. 1 confirm these calculated results.

Figure 17.8 shows the calculated directivity as a function of frequency. It is found that the reduction in the directivity from its maximum value across an analysis frequency range of 11.7–12 GHz is small. Note that the measured gain and return loss across this frequency range are presented in Ref. 1.

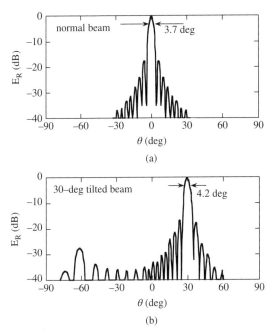

Figure 17.7 Radiation pattern (E_R pattern) at 11.85 GHz, where 396 helical elements are arrayed with an array spacing of $S_{rad} = 1.75$ cm. (a) Normal beam radiation. (b) 30°-tilted beam radiation. (Reproduced from Ref. [1] with permission from IEEE.)

Exercise

Determine the amplitude and phase of the helical element n shown in Fig. 17.4 [1,2,6].

Answer   First, we express the total electric field as a vector $\dot{E}$

$$\dot{E} = E_0 e^{j\delta_0}$$

$$= \sum_{m=1}^{N_a} E_m e^{j\delta_m} \tag{17.4}$$

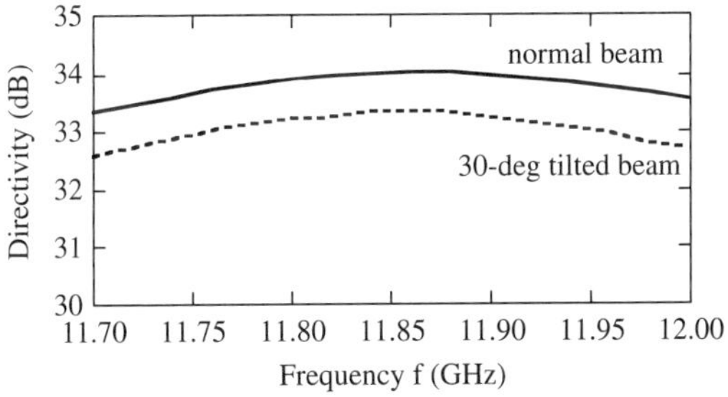

Figure 17.8 Directivity as a function of frequency. (Reproduced from Ref. [1] with permission from IEEE.)

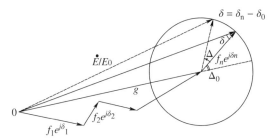

Figure 17.9   Radiation fields (normalized to E_0) from the array antenna elements. (Reproduced from Ref. [1] with permission from IEEE.).

where E_0 and δ_0 are the amplitude and the phase of the total electric field, respectively, and E_m and δ_m are the amplitude and the phase of the helical element m, respectively.

Second, we rotate the element n by Δ, as shown in Fig. 17.9. This changes the total electric field vector to

$$\dot{E} = E_0 e^{j\delta_0} - E_n e^{j\delta_n} + E_n e^{j(\delta_n+\Delta)} \tag{17.5}$$

Hence,

$$|\dot{E}|^2/E_0^2 = g^2 + f_n^2 + 2gf_n\cos(\Delta + \Delta_0) \tag{17.6}$$

where

$$f_n \equiv E_n/E_0 \tag{17.7}$$

$$\delta \equiv \delta_n - \delta_0 \tag{17.8}$$

$$g \equiv \left| \sum_{\substack{m=1 \\ m \neq n}}^{N_a} E_m e^{j\delta_m} \right| = \sqrt{(\cos\delta - f_n)^2 + (\sin\delta)^2} \tag{17.9}$$

$$\tan\Delta_0 = \sin\delta/(\cos\delta - f_n) \tag{17.10}$$

The maximum power of the total electric field appears at $\Delta = -\Delta_0$, and the ratio of the maximum power to the minimum power of the total electric field is written as

$$(g + f_n)^2/(g - f_n)^2 \equiv r^2 \tag{17.11}$$

Referring to Fig. 17.9, the relative amplitude f_n and phase $\delta = \delta_n - \delta_0$ for $g > f_n$ are

$$f_n = \Gamma/\sqrt{1 + 2\Gamma\cos\Delta_0 + \Gamma^2} \tag{17.12}$$

$$\delta = \tan^{-1}[\sin\Delta_0/(\cos\Delta_0 + \Gamma)] \tag{17.13}$$

where

$$\Gamma = f_n/g = (r - 1)/(r + 1) \tag{17.14}$$

Note that Δ_0 and Γ are obtained by rotating the helical element n around its axis and measuring the variation of the relative power $|\dot{E}|^2/E_0^2$.   ∎

17.3 APPLICATION EXAMPLES

Figure 17.10 presents two helical antenna arrays designed using the axis rotation technique and vector rotation method, where Fig. 17.10a shows a CP antenna for direct-broadcast-satellite reception (Ku-band operation) [7] and Fig. 17.10b shows a helical antenna array placed on a satellite for Mercury exploration [8]. In addition to these arrays, a two-layer helical antenna array for two-band operation has been realized, as shown in Fig. 17.11a, where the upper layer helical antenna array is used for a specific frequency band and the lower layer helical antenna array is used for a different frequency band. Figure 17.11b shows an application example of the two-layer helical antenna array [9,10], where the upper layer helical antenna array is designed as an X-band antenna (8.1–8.9 GHz) and the lower layer helical antenna array is designed as an S-band antenna (2.1–2.6 GHz), both acting as primary feed antennas for a VERA Cassegrain reflector [11]. The radiation patterns at X- and S-band frequencies are shown in Fig. 17.12, where a beam width of $22°$ for a $7\,dB$ reduced field intensity criterion is realized, meeting the requirement for both of the primary feed antennas. The parameters used for these primary feed antennas are summarized in Ref. 12.

(a)   (b)

Figure 17.10   Helical antenna arrays. (a) For direct-broadcast-satellite reception. (Courtesy of TDK Co.)
(b) For Mercury exploration (BepiColombo project). (Courtesy of JAXA.)

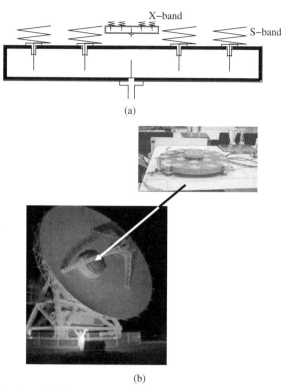

(b)

Figure 17.11   Two-layer helical antenna array for two-band operation. (a) Side view. (b) The primary feed for a VERA Cassegrain reflector.

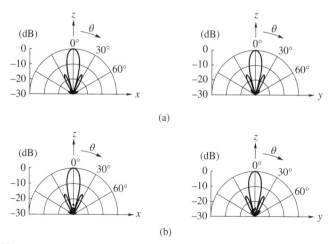

Figure 17.12   Radiation patterns for the two-layer primary-feed helical antenna array for a VERA Cassegrain reflector. (a) At 12.35 GHz. (b) At 8.55 GHz.

REFERENCES

1. H. Nakano, H. Takeda, Y. Kitamura, H. Mimaki, and J. Yamauchi, Low-profile helical array antenna fed from a radial waveguide. *IEEE Trans. Antennas Propag.*, vol. 40, no. 3, pp. 279–284, 1992.
2. S. Mano and T. Katagi, A method of measuring amplitude and phase of each radiating element of a phased array antenna. *Trans. IEICE Japan*, vol. 65, pp. 555–560, 1982.
3. M. Marcuvitz, *Waveguide Handbook*, New York: McGral-Hill, 1961.
4. F. J. Goebels and K. C. Kelly, Arbitrary polarization from annular slot planar antennas. *IRE Trans. Antennas Propag.*, vol. 9, pp. 342–349, 1961.
5. K. R. Carver, *A cavity-fed concentric ring phased array of helices for use in radio astronomy*, Ph.D. dissertation, Ohio state university, 1967.
6. H. Nakano, S. Matsumoto, and N. Ochiai, Improvement of the axial ratio for a circularly polarized multi-arm antenna. *Proc. Combined Conference of Four Institutions Related to Electrical Engineering*, Denki 4 Gakkai, Rengoutaikai, Hokuriku-shibu, Japan, 1975, p. B-2.
7. N. Misawa, H. Yamashita, K. Ishino, J. Yamauchi, H. Mimaki, and H. Nakano, Radiation characteristics of a low-profile helical array antenna with a radome. In: *Proc. Int. Sympo. Antennas Propag.*, ISAP, Sapporo, Japan, 1992, pp. 45–48.
8. http://www.isas.jaxa.jp/j/enterp/missions/mmo/index.shtml
9. H. Mimaki, H. Nakano, and T. Kasuga, Very compact S and X bands coaxial helical array feeds for VLBI antenna, *Proc. Asia-Pacific Radio Science Conference*, AP–RASCO'01, Tokyo, Japan, August 2001, p. 240.
10. T. Kasuga, H. Mimaki, and H. Nakano, Compact helical array antenna for VLBI co-axial S-X band feeds. *Proc. XXVIIth General Assembly of Int. URSI*, Maastricht, Germany, August 2002, pp. 2.1–2.4,
11. http://veraserver.mtk.nao.ac.jp/restricted/status05.pdf (Japanese).
12. L. Shafai, S. Sharma, and S. Rao, *Handbook of reflector antennas and feed systems*, vol. II (Feed systems), chapter 7. Boston: Artech House, 2013.

Chapter 18

Curl Antennas

The spiral antenna described in Chapter 16 is a CP (circularly polarized) nonresonant antenna that radiates a CP wave generated by a traveling wave current along the spiral arm. The helical antenna described in Chapter 17 is also a nonresonant CP antenna. The operational bandwidth of these CP antennas is wider than that of CP resonant antennas, such as CP patch and loop antennas having perturbation elements [1,2]. This chapter extends the discussion in Chapters 16 and 17 by presenting two array antennas, each using low-profile CP nonresonant radiation elements.

In Section 18.1, a high-gain normal-beam multicircular array antenna composed of nonresonant CP curl elements with a circular cavity is designed, adopting the same design techniques that are used in Chapter 17. The curl array element is excited through a vertical arm located at the antenna center, and is referred to as an internal-excitation curl [3–5]. In Section 18.2, a one-dimensional linear array antenna composed of tilted-beam curl elements is investigated, where each curl is excited through an arm connected to the outermost point of the curl, and is referred to as an external-excitation curl [6–8].

18.1 HIGH-GAIN NORMAL-BEAM ARRAY ANTENNA COMPOSED OF INTERNAL-EXCITATION CURL ELEMENTS

A low-profile internal-excitation curl element (antenna) is presented, which is a simplified version of a spiral antenna. This curl is used as an array element for a high-gain normal-beam array antenna. The discussion starts with the configuration of the internal-excitation curl and concludes with a curl array antenna design that uses a circular cavity as a feed system.

18.1.1 Internal-Excitation Curl Antenna as an Array Element

Figure 18.1 shows the configuration of a curl antenna. The antenna is made of a single wire of diameter $2a$ and a ground plane (GP). The antenna arm is composed

Low-Profile Natural and Metamaterial Antennas: Analysis Methods and Applications, First Edition.
Hisamatsu Nakano.
© 2016 The Institute of Electrical and Electronics Engineers, Inc. Published 2016 by John Wiley & Sons, Inc.

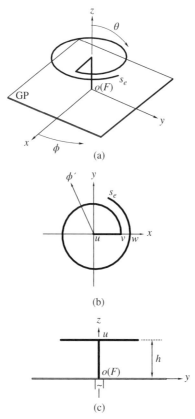

Figure 18.1   Configuration of a curl antenna. (a) Perspective view. (b) Top view. (c) Side view.
(Reproduced from Ref. [3] with permission from IEEE.)

of an inverted-L wire section and a curled wire section. The bottom end of the
inverted-L wire section is connected to the inner conductor of a coaxial feed line.
The radial distance from the center point u to a point on the curled wire, r, is
defined by the Archimedean function:

$$r = a_{sp}\phi'$$ (18.1)

where a_{sp} is the spiral constant and ϕ' is the winding angle starting at ϕ_{st} and ending at ϕ_{end}. The height of the curled wire section above the ground plane, h, is
chosen to be small, compared with the wavelength at the design frequency.

The curled wire section is designed to radiate a CP beam in the z-direction. For
this, the following two conditions must be satisfied: (1) the curled wire section has
an annular region whose circumference is one wavelength (1λ) at the design frequency, and (2) the curled wire arm has sufficient length for the current traveling
toward the arm end to decay gradually so that the current reflected from the arm
end is as small as possible.

18.1.2 Radiation from an Internal-Excitation Curl Antenna

The arm end angle ϕ_{end} for the curl is determined by taking into account the axial ratio at the design frequency 11.85 GHz (wavelength $\lambda_{11.85} = 25.32$ mm). Figure 18.2 shows the axial ratio as a function of ϕ_{end} with antenna height h as a parameter, where the ground plane is assumed to be of infinite extent and the following parameters are used: $a = 0.3$ mm $= 0.01185\lambda_{11.85}$, $a_{sp} = 0.18$ mm/rad $= 0.00711\lambda_{11.85}$/rad, and $\phi_{st} = 6\pi$ rad (note that the curled section starts at $r = 0.134$ $\lambda_{11.85}$). It is found that the best axial ratio is obtained when $(h, \phi_{end}) = (3.8$ mm $= 0.15\lambda_{11.85}$, 26.3 rad$) \equiv (h_{op}, \phi_{op})$. The curled wire section in this case has a one-wavelength annular region, meeting condition (1). Figure 18.3 shows the current distribution $I = I_r + jI_i$ at 11.85 GHz along the antenna arm specified by (h_{op}, ϕ_{op}). It is found that the current flows in a traveling wave fashion with decay. This meets condition (2). Therefore, parameters (h_{op}, ϕ_{op}) that meet both conditions (1) and (2) are chosen for the following discussion.

The current distribution depicted in Fig. 18.3 generates a CP wave, as shown in Fig. 18.4, where E_R is the right-handed CP component (copolarization component) and E_L is the left-handed CP component (cross-polarization component). It is found that the asymmetric antenna structure with respect to the z-axis makes the radiation pattern quasi-symmetric. However, this quasi-symmetric radiation does not become a serious problem in forming a high-gain beam, by virtue of the fact that the radiation pattern of the curl is wide around the z-axis. This is verified in Section 18.1.3. Note that the radiation pattern for E_R, which is used later for the directivity calculation of an array antenna, is approximated to be $[(\cos(0.571\theta)]^5$.

The results for the current and radiation pattern shown in Figs. 18.3 and 18.4 are for a frequency of 11.85 GHz. As long as the curl supports a smoothly decaying current, it is expected that the curl will maintain stable radiation characteristics across a wide-frequency range. This is confirmed by the frequency response of the radiation characteristics around 11.85 GHz, such as the axial ratio (Fig. 18.5) and the gain (Fig. 18.6). The bandwidth for a 3 dB axial ratio criterion is 6.7%, across

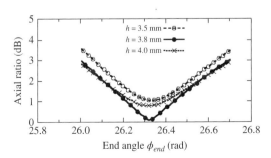

Figure 18.2 Axial ratio as a function of ϕ_{end} at a design frequency of 11.85 GHz, with antenna height h as a parameter. (Reproduced from Ref. [3] with permission from IEEE.)

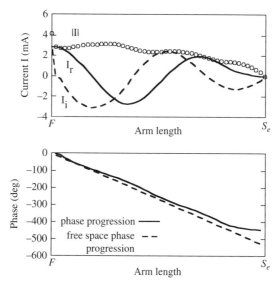

Figure 18.3 Current distribution, $I = I_r + jI_i$, along the antenna arm at 11.85 GHz, where parameters $(h, \phi_{end}) = (3.8 \text{ mm} = 0.15\lambda_{11.85}, 26.3 \text{ rad}) \equiv (h_{op}, \phi_{op})$ are used. (Reproduced from Ref. [3] with permission from IEEE.)

which the half-power beam width is approximately 73°. The gain has a wideband characteristic with a value of approximately 8.4 dBi, which is comparable to the gain of the low-profile helical antenna in Chapter 17. Note that the frequency response of the axial ratio and the gain are validated by experimental work in Ref. 3.

18.1.3 Internal-Excitation Curl Array Antenna

The goal of this section is to form a high-gain normal beam. For this, N_a curl antenna elements are arrayed, as shown in Fig. 18.7, where each vertical probe of the inverted-L wire section is inserted into a circular cavity. As in the helical array antenna discussed in Chapter 17, the cavity depth (height) is chosen to be less than one-half the wavelength at the design frequency (11.85 GHz); the number of curl elements on the pth circumference (ring) is $6p$; and the spacing between array elements is given by $S_{cir} = \pi S_{rad}/3$.

Figure 18.8 shows the radiation pattern for an array antenna composed of 168 curl elements (six rings) at 11.85 GHz, where a spacing of $S_{rad} = 2 \text{ cm} = 0.79\lambda_{11.85}$ is used. The half-power beam width obtained using the principle of pattern multiplication [9] is approximately 4.8°, which agrees with the measured beam width. Note that the amplitude and phase for the array elements are adjusted using the axis rotation technique [10] and the vector rotation method [11]. Figure 18.9 shows the measured frequency response of the gain together with the aperture efficiency η. It is found that the aperture efficiency reaches a maximum of more than 90%.

E_R ——
E_L -----

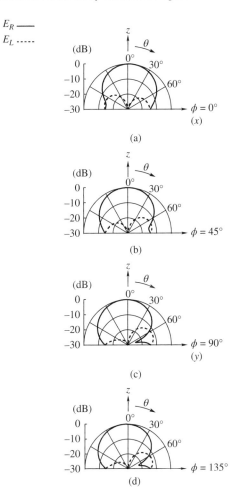

(a)

(b)

(c)

(d)

Figure 18.4 Radiation pattern at 11.85 GHz, where parameters $(h, \phi_{end}) = (3.8\,\text{mm} = 0.15\lambda_{11.85}, 26.3\,\text{rad}) \equiv (h_{op}, \phi_{op})$ are used. (a) In the elevation plane at $\phi = 0°$. (b) In the elevation plane at $\phi = 45°$. (c) In the elevation plane at $\phi = 90°$. (d) In the elevation plane at $\phi = 135°$. (Reproduced from Ref. [3] with permission from IEEE.)

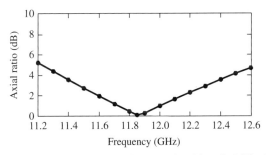

Figure 18.5 Frequency response of the axial ratio. (Reproduced from Ref. [3] with permission from IEEE.)

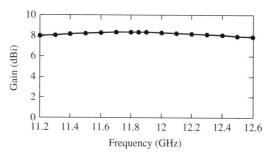

Figure 18.6   Frequency response of the gain. (Reproduced from Ref. [3] with permission from IEEE.)

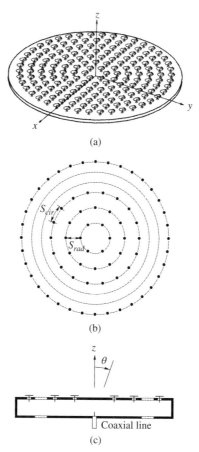

Figure 18.7   High-gain normal beam array antenna composed of internal-excitation curl antenna elements. (a) Perspective view. (b) Top view. (c) Side view. (Reproduced from Ref. [3] with permission from IEEE.)

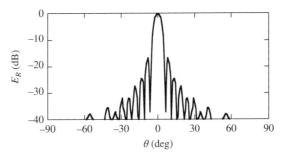

Figure 18.8 Radiation pattern for an array composed of 168 curl elements at 11.85 GHz. (Reproduced from Ref. [3] with permission from IEEE.)

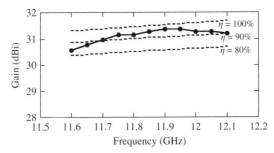

Figure 18.9 Frequency response of the gain, together with the aperture efficiency η. (Reproduced from Ref. [3] with permission from IEEE.)

18.2 HIGH-GAIN TILTED-BEAM ARRAY ANTENNA COMPOSED OF EXTERNAL-EXCITATION CURL ELEMENTS

When radiation elements (each radiates a tilted CP beam) are arrayed side-by-side in a row, this one-dimensional linear array forms a high-gain CP-tilted beam. The one-dimensional linear array has the advantage that the installation area required for the array is smaller than that of conventional two-dimensional phased arrays using nontilted CP beam elements. This section starts with the realization of an element radiating a CP-tilted beam and concludes with a discussion of the gain of a one-dimensional CP linear array.

18.2.1 External-Excitation Curl Antenna as an Array Element

The angular separation between the ground and a satellite is assumed to be approximately 30° and hence the direction of the beam radiated from the CP curl element/ antenna to be designed here is selected to be $\theta = 30°$, measured from the zenith. The following three requirements are imposed on the antenna design: (1) a small antenna circumference not exceeding two wavelengths (2λ), that is, a small antenna diameter not exceeding $2\lambda/\pi = 0.64$ wavelength, which accommodates suppression of grating lobes when the antenna is used as an array element, (2) a low-profile structure not exceeding half the wavelength, and (3) a robust structure. Meeting these requirements contribute to realizing a robust, low-profile antenna with small dimensions.

A center-fed single-arm spiral antenna backed by a conducting reflector radiates a CP beam (whose maximum radiation appears near the antenna axis normal to the spiral plane) when the antenna circumference is between one wavelength and two wavelengths. As the antenna circumference is increased from two wavelengths, the CP radiation beam becomes tilted off the antenna axis, as discussed in Chapter 16. However, this structure does not meet requirement (1).

A helical antenna is also a CP antenna. When the circumference of the helical antenna is between 3/4 and 4/3 wavelengths, the antenna radiates an axial CP beam [9]. As the antenna circumference is further increased from 4/3 wavelengths, the radiation becomes tilted. For this tilted beam to be circularly polarized, the antenna height above the ground plane (longitudinal length) must be greater than one wavelength. It follows that this helical antenna does not meet requirement (2).

Figure 18.10 shows an antenna structure that achieves requirements (1), (2), and (3). Features of the structure are a curled strip line (on the top surface of a hollow dielectric cylinder of relative permittivity ε_r and thickness t) and a different curved strip line (on the side surface of the cylinder). The latter strip line (of width W_{ex}) is used to excite the outermost point of the curl, P, and designated as a *wound excitation line* (WD-line). Note that a small segment of length $\Delta h = 0.6\,\text{mm}$ is added to the bottom of the WD-line in order to facilitate soldering the WD-line to the inner conductor of a coaxial feed line. The WD-line, having radius r_{max}, is wound with an angle of ϕ_{WD} around the z-axis (ϕ_{WD} is called the *rotation angle*,

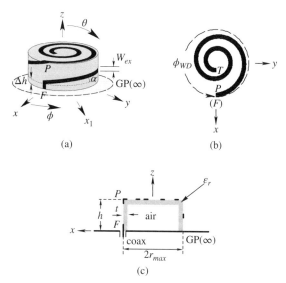

Figure 18.10   External-excitation curl antenna. (a) Perspective view. (b) Top view. (c) Side view. (Reproduced from Ref. [6] with permission from IEEE.)

measured from the x-axis to a projection of the outermost point P onto the x–y plane), with a slope of angle α. Note that the winding sense of the WD-line is the same as that of the curl in order to strengthen the principal radiation (copolarization) component E_R, resulting in a decrease in the cross-polarization component E_L.

The radial distance from the center point of the top surface [at height h (denoted as o')] to a point on the curl, r_{CL}, is defined by an equiangular function:

$$r_{CL} = r_{max}e^{-a(\phi'-\phi_{WD})} \qquad (18.2)$$

where r_{max} is the maximum radial distance of the curl (antenna radius), measured from the center point o'; a is the spiral constant; angle ϕ' starts at ϕ_{WD} and ends at $\phi' = \phi_{WD} + \phi_{CL}$, with ϕ_{CL} being the winding angle of the curl. Note that requirement (1) is met by setting r_{max} appropriately. The rotation angle ϕ_{WD} and slope α are chosen such that the antenna height meets requirement (2). The WD-line is printed on a hollow dielectric cylinder, meeting requirement (3).

We use a frequency range from 11.7 GHz ($\equiv f_L$) to 12.75 GHz ($\equiv f_H$) as the external-excitation curl antenna design frequency range. The total radiation field is the sum of the fields from the curled strip section and the WD-line. Figure 18.11 shows the radiation pattern at a design center frequency of $f_c = 12.225$ GHz with slope α as a parameter, where the following configuration parameters are used together with the parameters shown in Table 18.1: curl winding angle $\phi_{CL} = 4\pi$ rad $\equiv \phi_{CL\text{-}0}$, maximum radius $r_{max} = 6.8$ mm $\equiv r_{max\text{-}0}$, and rotation angle $\phi_{WD} = 2\pi$ rad $\equiv \phi_{WD\text{-}0}$. This configuration is expressed as $(\alpha, \phi_{CL}, r_{max}, \phi_{WD}) = $ (variable, $\phi_{CL\text{-}0}, r_{max\text{-}0}, \phi_{WD\text{-}0}$). It is found that slope α affects the beam direction angle θ_{max} in the elevation plane [$\theta_{max} = 29°$ for $\alpha = 4°$ and $\theta_{max} = 39°$ for $\alpha = 12°$].

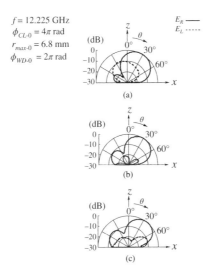

Figure 18.11 Radiation pattern at $f_c = 12.225$ GHz, with slope α as a parameter. (a) $\alpha = 4°$. (b) $\alpha = 8°$. (c) $\alpha = 12°$. (Reproduced from Ref. [6] with permission from IEEE.)

Table 18.1   Basic Parameters

Symbol	Value
a	0.105 mm/rad
W_{ex}	0.8 mm
t	1.0 mm
ε_r	2.0
Δh	0.6 mm

Source: Reproduced from Ref. 6 with permission
from IEEE.

It is also found that an appropriate selection of slope α leads to a small cross-polarization component E_L that is attributed to the fact that the outgoing current flowing toward the arm end T (the innermost point of the curl) is dominant, compared with the incoming current flowing toward the feed point F (the reflected current from the arm end T). For a better understanding of this fact, the current distribution for $\alpha = 4°$ and $\alpha = 8°$ is shown in Fig. 18.12, where the current ($I = I_r + jI_i$ with $|I|$ being the amplitude) is calculated by integrating the magnetic field around the antenna arm, which is obtained using FDTDM analysis. The current for $\alpha = 8°$ ($\equiv \alpha_0$) exhibits smoother decay from feed point F than the current for $\alpha = 4°$;

$f = 12.225$ GHz

$\phi_{CL-0} = 4\pi$ rad

$r_{max-0} = 6.8$ mm

$\phi_{WD-0} = 2\pi$ rad

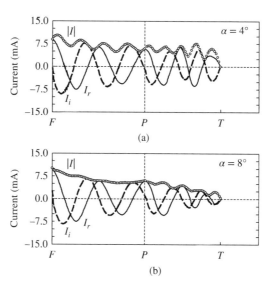

Figure 18.12   Current distribution $I = I_r + jI_i$, where $|I|$ is the amplitude. (a) $\alpha = 4°$. (b) $\alpha = 8°$ ($\equiv \alpha_0$). (Reproduced from Ref. [6] with permission from IEEE.)

Table 18.2   Fixed Parameters

Symbol	Value
$r_{max\text{-}0}$	6.8 mm
$\phi_{CL\text{-}0}$	4π rad
$\phi_{WD\text{-}0}$	2π rad
α_0	$8.0°$

Source: Reproduced from Ref. 6 with permission from IEEE.

the current for $\alpha = 4°$ has ripples in the amplitude, $|I|$, due to the presence of the incoming current flowing toward feed point F.

Note that the antenna height h (including Δh) is approximately 0.15 wavelength at $\alpha = 4°$ and 0.28 wavelength at $\alpha = 8°$ at a design frequency of $f_H = 12.75$ GHz, meeting requirement (2). Other parametric studies of ϕ_{CL}, r_{max}, and ϕ_{WD} are found in Ref. 6. Based on these studies, the parameters shown in Table 18.2 are fixed for the following discussion.

The input impedance across the design frequency range does not match a 50 Ω feed line. To realize input impedance matching and obtaining a small VSWR, the antenna arm near the antenna input F is transformed, as shown in Fig. 18.13a, where the transformed section (matching section) is expressed by length L_m and width W_m. Figure 18.13b shows the frequency response of the VSWR for $(L_m, W_m) = (5.2\,\text{mm}, 2.4\,\text{mm})$. It is found that this simple arm transformation provides a sufficiently wide bandwidth, exceeding the design frequency range. Note that the length L_m is approximately $\lambda_g/4$ at the design center frequency, where λ_g is calculated using $\lambda/[(\varepsilon_r + 1)/2]^{1/2}$ with λ being the free-space wavelength.

Figure 18.13c shows a fabricated external-excitation curl, where the ground plane is circular and large relative to the wavelength of the design center frequency $(f_c = 12.225\,\text{GHz})$, having a diameter of $D_{GP} = 200\,\text{mm} = 8.15\lambda_{12.225}$, where $\lambda_{12.225}$ is the wavelength at 12.225 GHz. The measured antenna characteristics for this fabricated antenna are found in Ref. 6.

18.2.2 External-Excitation Curl Array Antenna

Figure 18.14 shows two curl elements separated by distance d_x on the x-axis, where the ground plane GP is assumed to be of infinite extent [7]. Note that the configuration parameters for each curl element are the same as those shown in Tables 18.1 and 18.2. Also, note that each element is rotated around its axis (cylindrical axis normal to the GP) so that the maximum radiation is in the y–z plane.

Figure 18.15 depicts the $|S_{21}|$ characteristic at the design center frequency $(f_c = 12.225\,\text{GHz})$ as a function of the element spacing d_x. It is found that $|S_{21}|$ is less than $-30\,\text{dB}$ when $d_x \geq 0.8\lambda_{12.225}$ [7]. Based on this result, a spacing of $d_x = 0.8\lambda_{12.225}$ is used for the following N-element array antenna, shown in Fig. 18.16.

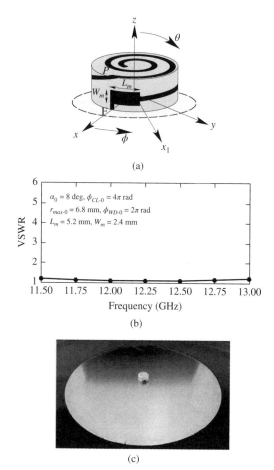

(a)

(b)

(c)

Figure 18.13   Impedance matching. (a) Arm transformation. (b) Frequency response of the VSWR. (c) Fabricated external-excitation curl antenna. (Reproduced from Ref. [6] with permission from IEEE.)

Figure 18.17 shows the gain for a right-handed CP wave as a function of the number of curl elements N. This gain is the gain in the tilted beam direction and is enhanced with increasing N, by virtue of a reduction in mutual effects; a gain of approximately 22 dBi is obtained when $N = 16$ [7]. The radiation pattern

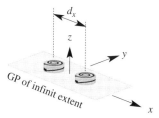

Figure 18.14   Two curl elements on the x-axis. (Reproduced from Ref. [7] with permission from IEICE.)

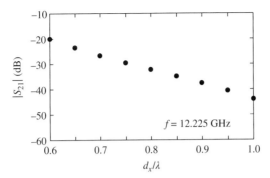

Figure 18.15   Scattering parameter $|S_{21}|$. (Reproduced from Ref. [7] with permission from IEICE.)

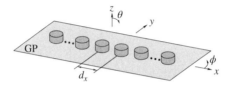

Figure 18.16   One-dimensional linear array antenna composed of N external-excitation curl elements.

when $N = 16$ is illustrated in Fig. 18.18. It is revealed that the copolarization component E_R in the elevation plane, depicted in Fig. 18.18a, has almost the same pattern as that of a single curl antenna, shown in Fig. 18.11b, where the beam is in the $\theta \approx 30°$ direction. It is also revealed that the E_R pattern in the $\theta_{max} - \phi$ plane [$(\theta, \phi) = (30°,$ varied) plane] is narrowed by array effects, as shown in Fig. 18.18b, leading to a high gain. Note that the radiation pattern and gain are not sensitive to a change in frequency ranging from $f_L = 11.7\,\text{GHz}$ to $f_H = 12.75\,\text{GHz}$ (design frequency range).

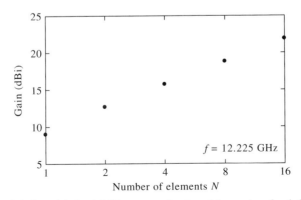

Figure 18.17   Gain for a right-handed CP wave as a function of the number of curl elements N. (Reproduced from Ref. [7] with permission from IEICE.)

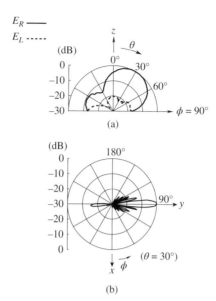

Figure 18.18   Radiation pattern when $N = 16$. (a) In the elevation plane. (b) In the $\theta_{max} - \phi$ plane. (Reproduced from Ref. [7] with permission from IEICE.)

REFERENCES

1. H. Nakano, K. Vichien, T. Sugiura, and J. Yamauchi, Singly-fed patch antenna radiating a circularly polarized conical beam. *Electron. Lett.*, vol. 26, no. 10, pp. 638–640, 1990.
2. H. Nakano, A numerical approach to line antennas printed on dielectric materials. *Comput. Phys. Commun.*, vol. 68, pp. 441–450, 1991.
3. H. Nakano, S. Okuzawa, K. Ohishi, M. Mimaki, and J. Yamauchi, A curl antenna. *IEEE Trans. Antennas Propag.*, vol. 41, no. 11, pp. 1570–1575, 1993.
4. H. Nakano, M. Yamazaki, and J. Yamauchi, Electromagnetically coupled curl antennas. *Electron. lett.*, vol. 33, no. 12, pp. 1003–1004, 1997.
5. H. Nakano and H. Mimaki, Axial ratio of a curl antenna. *Proc. IEEE Microw. Antennas Propag.*, vol. 144, no. 6, pp. 488–490, 1997.
6. H. Nakano, S. Kirita, M. Mizobe, and J. Yamauchi, External-excitation curl antenna. *IEEE Trans. Antennas Propag.*, vol. 59, no. 11, pp. 3969–3977, 2011.
7. S. Kirita, J. Yamauchi, and H. Nakano, Small helical-spiral antenna. *Proc. IEICE Soci. Conf.*, 2010, p. B-1-112.
8. K. Hirose, T. Wada, and H. Nakano, An outer-fed spiral antenna array radiating a circularly polarized conical beams. *Electromagnetics*, vol. 20, no. 4, pp. 295–310, 2000.
9. J. Kraus, R. Marhefka, *Antennas*, 3rd edn, NY: McGraw Hill, 2002.
10. H. Nakano, S. Matsumoto, and N. Ochiai, Improvement of the axial ratio for a circularly polarized multi-arm antenna. In: *Proc. Combined Conference of Four Institutions Related to Electrical Engineering* (Denki 4 Gakkai, Rengoutaikai, Hokuriku-shibu, Japan), 1975, p. B-2.
11. S. Mano and T. Katagi, A method for measuring amplitude and phase of each radiating element of a phased array antenna. *Trans. IEICE Japan*, vol. J-65-B, pp. 555–560, 1982.

Part III

Low-Profile Metamaterial Antennas

Chapter 19

Metaline Antenna

The radiation from a leaky wave antenna, where the wave velocity v (phase velocity) is faster than the velocity of light c, forms a beam inclined at an angle $\theta = \sin^{-1} c/v$ ($0° < \theta < 90°$), which is called a forward inclined radiation beam. In this chapter, we design a line antenna where the beam scans seamlessly with change in frequency from the backward direction ($\theta = -90°$) to the forward direction ($\theta = 90°$) through the broadside direction ($\theta = 0°$). This line antenna is designated as a meta-material-based line antenna or metaline. The design is performed using transmission line theory [1].

19.1 UNIT CELL

Figure 19.1 shows the unit cell of an infinitely long parallel transmission line. The cell has a length p [m], which is short relative to the operating wavelength. L' [H/m] and C' [F/m] are, respectively, the inductance and the capacitance inherently distributed along the line, and C_Z [F] and L_Y [H] are, respectively, a capacitance and an inductance added to the cell. The series impedance, composed of L' and C_Z, that is, $Z' = j\omega L' + \frac{1}{j\omega(pC_Z)}$ (Ω/m), resonates at an angular frequency of

$$\omega_Z = \frac{1}{\sqrt{L'C_Z'}} \tag{19.1}$$

where $C_Z' = pC_Z[F \cdot m]$. The shunt admittance, composed of C' and L_Y, that is, $Y' = j\omega C' + \frac{1}{j\omega(pL_Y)}$ (S/m), resonates at an angular frequency of

$$\omega_Y = \frac{1}{\sqrt{L_Y'C'}} \tag{19.2}$$

where $L_Y' = pL_Y[H \cdot m]$. When these resonance angular frequencies for the series impedance Z and shunt admittance Y occur at the same angular frequency ω_T, that is,

$$\omega_Z = \omega_Y = \omega_T \tag{19.3}$$

Low-Profile Natural and Metamaterial Antennas: Analysis Methods and Applications, First Edition.
Hisamatsu Nakano.
© 2016 The Institute of Electrical and Electronics Engineers, Inc. Published 2016 by John Wiley & Sons, Inc.

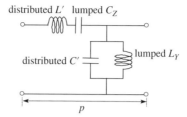

distributed L' lumped C_Z

distributed C' lumped L_Y

p

Figure 19.1 Unit cell of a transmission line.

this $\omega_T = 2\pi f_T$ is designated as the *transition angular frequency*, or simply the *transition frequency*.

19.2   NATURAL CHARACTERISTIC IMPEDANCE Z_{NTR}, BLOCH IMPEDANCE Z_B, AND PHASE CONSTANT β

To realize the unit cell shown in Fig. 19.1, a microstrip line of width w, printed on a dielectric substrate of relative permittivity ε_r and thickness B, is prepared as a base line (see Fig. 19.2). The characteristic impedance (*natural* characteristic impedance Z_{NTR}) of this line is defined by

$$L' = C'(Z_{NTR})^2 \tag{19.4}$$

The transmission line shown in Fig. 19.3 is made by subdividing the base line in Fig. 19.2 into numerous strip segments, each having length p_0. The cell length is defined as $p = 2(p_0 + \Delta g)$, where Δg is the gap between neighboring strip segments. The total metaline length is given by pM, where M is the total number of cells. A vertical conducting pin (via) of length h_{via} and radius r_{via} is connected to the central strip segment of the unit cell and is shorted to the ground plane through inductance L_Y. Neighboring strip segments are connected through capacitance $2C_Z$. This line is excited from point F and shorted to the ground plane $(GP_x \times GP_y)$ from point T to resistive load R_B, and called a metaline (or CRLHTL [1]). Note that values for L_Y and $2C_Z$ in practice are determined taking into account the effects of the pin and gap, respectively.

The Bloch impedance Z_B of this metaline is given as [2]

$$Z_B = \pm \frac{B}{\sqrt{A^2 - 1}} = \frac{\pm\sqrt{D^2 - 1}}{C} \tag{19.5}$$

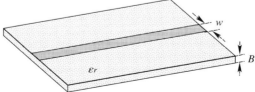

w

ε_r

B

Figure 19.2 Microstrip transmission line as a base line.

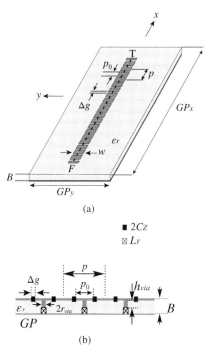

(a)

■ $2C_Z$
⊠ L_Y

(b)

Figure 19.3   Metaline. (a) Perspective view. (b) Side view.

where A, B, C, and D are F-matrix elements, which are related to the scattering matrix (S-matrix) elements. The propagation constant when the attenuation across the unit cell is negligible is expressed by $\gamma = j\beta$, where

$$\beta = \frac{1}{p}\cos^{-1}\left(\frac{A+D}{2}\right) \tag{19.6}$$

Exercise

Depict the Bloch impedance, dispersion diagram, and radiation pattern for a metaline antenna composed 20 cells ($M = 20$), using values shown in Tables 19.1 and 19.2 and $R_B = 80\,\Omega$.

Answer   Figure 19.4 shows the Bloch impedance as a function of frequency. The Bloch impedance shows a nearly constant value at frequencies around the transition frequency $f_T = 3\,\text{GHz}$.

Table 19.1   Parameters for a Microstrip Line

Symbol	Value
w	2 mm
B	1.6 mm
ε_r	2.6

Table 19.2   Parameters for a Metaline Antenna

Symbol	Value	Symbol	Value	Symbol	Value
p	10 mm	r_{via}	0.5 mm	$2C_Z$	1.06 pF
p_0	4 mm	h_{via}	0.6 mm	L_Y	3.83 nH
Δg	1.0 mm	pC'	0.61 pF	GP_x	240 mm
f_T	3.0 GHz	pL'	3.79 nH	GP_y	50 mm

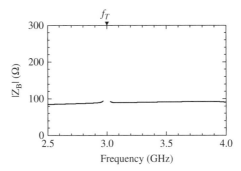

Figure 19.4   Frequency response of the Bloch impedance.

Figure 19.5 depicts the dispersion diagram for the unit cell, where k_0 is the phase constant in free space, and frequencies f_L and f_U are the lower- and upper-edge frequencies for a fast wave, respectively. It is clearly seen that the phase constant is negative below the transition frequency f_T and positive above the transition frequency.

Figure 19.6 shows the frequency response of the radiation pattern. It is found that the metaline forms backward inclined radiation at frequencies below the transition frequency f_T. The backward inclined radiation occurs due to the fact that the current flows with a progressive phase constant toward the arm end T; in other words, the phase delays toward the feed point F [note: conventional natural line antennas (leaky wave antennas) do not have such a characteristic]. At f_T, the radiation forms a broadside beam (beam in the z-axis direction). As the frequency is increased past f_T, the radiation forms a forward inclined beam, because the

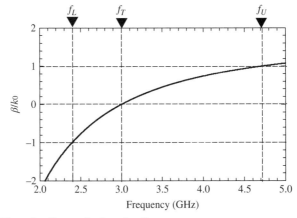

Figure 19.5   Dispersion diagram for the unit cell.

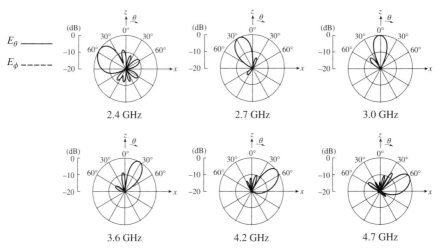

Figure 19.6   Frequency response of the radiation pattern for a 20-cell metaline antenna.

phase constant of the current is positive and hence the phase delay is in the positive x-direction. ■

19.3 TWO-METALINE ANTENNAS

The metaline shown in Fig. 19.3 is designed as a beam-scanning antenna, where a broadside beam is obtained across a very narrow frequency region centered at the transition frequency f_T. Figure 19.7 shows a two-metaline antenna for obtaining a

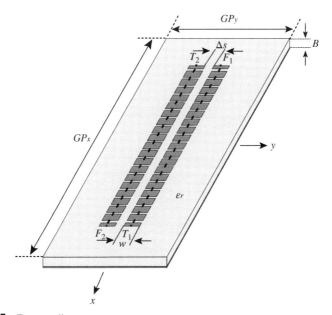

Figure 19.7   Two-metaline antenna.

Table 19.3 Parameters for a Two-Metaline Antenna

Symbol	Value	Symbol	Value	Symbol	Value
w	8.8 mm	r_{via}	0.5 mm	$2C_Z$	1.02 pF
B	3.2 mm	h_{via}	2.2 mm	L_Y	1.28 nH
ε_r	2.6	GP_x	120 mm	R_B	50 Ω
p_0	4.0 mm	GP_y	50 mm	f_T	3.0 GHz
Δg	1.0 mm	Δs	4.0 mm		
p	10 mm				

broadside beam across a wide frequency region around f_T, where points F_1 and F_2 are feed points and points T_1 and T_2 are shorted to the ground plane through resistive loads. The amplitude of the excitation at each of the points F_1 and F_2 is the same, with an excitation phase difference of 180°. This excitation provides constructive addition of the fields radiated from the two metalines at frequencies around f_T [3,4].

Exercise

Reveal the frequency response of the radiation pattern, gain, and VSWR for the two-metaline antenna shown in Fig. 19.7, specified by Table 19.3.

Answer Figure 19.8 shows the frequency response of the radiation pattern across a frequency bandwidth of 40%. The reduction in the gain from its maximum across this bandwidth is less than 3 dB, as shown in Fig. 19.9, and the VSWR is less than 2, as shown in Fig. 19.10. G_θ and G_ϕ in Fig. 19.9 show the gains for x- and y-polarized waves, respectively.

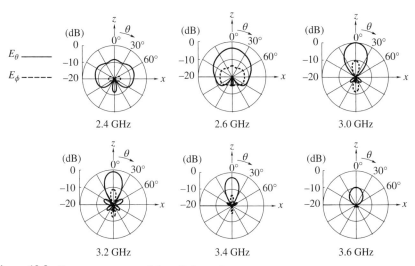

Figure 19.8 Frequency response of the radiation pattern.

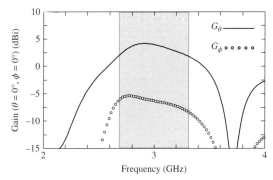

Figure 19.9 Frequency response of the gain.

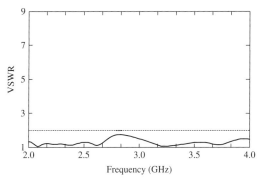

Figure 19.10 Frequency response of the VSWR.

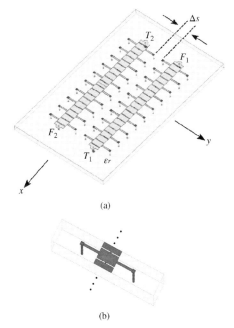

Figure 19.11 Shorted stubs realizing L_Y. (a) Two-metaline antenna with shorted stubs. (b) Antenna unit with two stubs.

It follows that the two-metaline antenna has a wideband characteristic with respect to the broadside pattern, gain, and VSWR [3,4]. Note that the inductors L_Y for the metalines can be realized by stubs shorted to the ground plane, as shown in Fig. 19.11. However, the distance between the two metalines Δs becomes larger than that for the structure shown in Fig. 19.7, making the antenna less compact. ■

REFERENCES

1. C. Caloz and T. Itoh, *Electromagnetic metamaterials*, NJ: Wiley, 2006.
2. R. Collin, *Foundations for microwave engineering*, NY: McGraw, 1966.
3. K. Sakata, J. Yamauchi, and H. Nakano, Double metaline antenna. In: *Proc. IEICE General Conference*, Shiga, March 2015, p. B-1-93.
4. K. Sakata, J. Yamauchi, and H. Nakano, Double metaline antenna with a dielectric slab. In: *Proc. IEICE Society Conference*, Miyagi, September 2015, p. B-1-111.

Chapter 20

Metaloop Antenna for Linearly Polarized Radiation

$\mathbf{A}$ natural loop antenna (LA) is a fundamental radiation element, just like the natural dipole antenna. When the natural LA is located in free space and the circumference of the loop is one wavelength (1λ), the radiation is linearly polarized (LP) with a maximal intensity in the $\pm z$-direction normal to the LA plane. In most applications, this bidirectional radiation is transformed into unidirectional radiation to enhance transmission efficiency. Placing a conducting plate (reflector) behind the LA is a practical and simple technique for obtaining unidirectional radiation. In such a case, the distance between the LA and the conducting plate is chosen to be one-quarter wavelength ($\lambda/4$) at the operating frequency so that the wave reflected from the conducting plate is constructively superimposed onto the direct radiation from the LA into free space.

This chapter presents a metamaterial-based loop antenna (metaloop) for unidirectional radiation [1,2], which differs from the natural LAs. The metaloop has an antenna height of less than 0.03λ, much smaller than the antenna height of the natural LAs. The radiation characteristics of the metaloop are analyzed and discussed.

20.1 METALOOP CONFIGURATION

The conducting arm of the metaloop in Fig. 20.1a is printed on a dielectric substrate of relative permittivity ε_r and thickness B. As in the straight metaline antenna in Chapter 19, numerous strip segments (each having length p_0 and width w) constitute the antenna arm, where neighboring strip segments are separated by gap Δg, as shown in Fig. 20.1c. The iterative region of length $p = 2(p_0 + \Delta g)$ is defined as the unit cell.

The ground plane (GP) has rectangular slits, each specified by a_w, a_p, i_w, and i_p, as shown in Fig. 20.1d, where area $i_p \times i_w$ is called an *island*. To obtain a

Low-Profile Natural and Metamaterial Antennas: Analysis Methods and Applications, First Edition.
Hisamatsu Nakano.
© 2016 The Institute of Electrical and Electronics Engineers, Inc. Published 2016 by John Wiley & Sons, Inc.

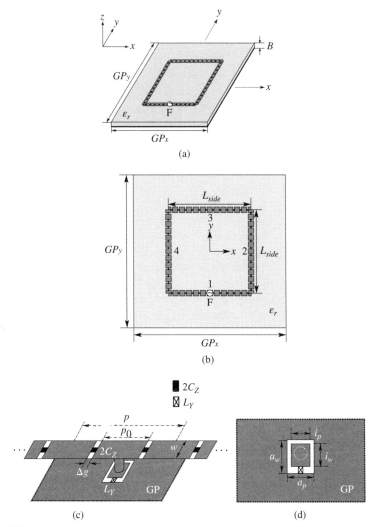

Figure 20.1 Metaloop antenna arm. (a) Perspective view. (b) Top view. (c) Exploded view of the antenna arm. (d) Island within the ground plane GP. (Reproduced from Ref. [1] with permission from IEEE.)

metamaterial property (a left-handed property: $\beta < 0$), the center strip segment of the unit cell is connected to the island through a conducting pin (via) of radius r_{via} and the island is shorted to the ground plane through inductance L_Y (inserted into the slit between the island and the ground plane). Neighboring strip segments are connected through capacitance $2C_Z$. The cell shown in Fig. 20.1c and d is designated as the island cell.

The metaloop is excited from point F in balanced mode (equal amplitude and $180°$ phase difference). Figure 20.2 depicts the dispersion diagram for the unit cell,

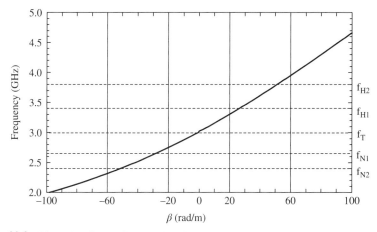

Figure 20.2   Dispersion diagram for the unit cell.

where the transition frequency is chosen to be $f_T = 3$ GHz, and the parameters used are shown in Table 20.1. The peripheral length of the arm corresponds to one guided wavelength ($1\lambda_g$) at frequencies f_{N1} and f_{H1}, and $2\lambda_g$ at frequencies f_{N2} and f_{H2}.

20.2 SINGLE- AND DUAL-PEAK BEAMS

Figure 20.3 shows the radiation patterns in the x–z plane, where E_θ and E_ϕ, respectively denote the radiation field components in the θ-direction and ϕ-direction for the spherical coordinate system. The radiation patterns at frequencies f_{N1} and f_{N2} are formed by currents with different *negative* phase constants, β_{N1} and β_{N2}, respectively; in contrast, the radiation patterns at frequencies f_{H1} and f_{H2} are formed by currents with different *positive* phase constants, β_{H1} and β_{H2}, respectively. It is found that single-peak radiation (broadside radiation) is obtained at f_{N1} and f_{H1}, where the copolarization component is E_θ in the x–z plane (and E_ϕ in the y–z plane). It is also found that dual-peak radiation is obtained at f_{N2} and f_{H2}, where the copolarization component of the radiation is E_ϕ in the x–z plane (and E_ϕ in the y–z plane). The radiation pattern is symmetric with respect to the z-axis in the x–z plane

Table 20.1   Parameters for the Metaloop Antenna

Symbol	Value	Symbol	Value (mm)	Symbol	Value (mm)	Symbol	Value
w	4.0 mm	Δg	0.5	GP_y	110	a_w	2.45 mm
B	3.2 mm	p	10	L_{side}	60	a_p	2.00 mm
ε_r	2.6	r_{via}	0.5	i_w	1.65	$2C_Z$	1.1 pF
p_0	4.5 mm	GP_x	110	i_p	1.40	L_Y	2.5 nH

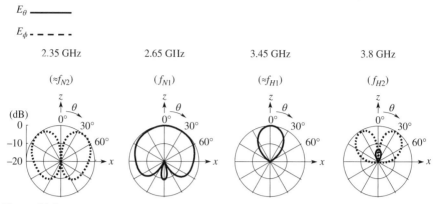

Figure 20.3 Radiation patterns for the metaloop.

and asymmetric with respect to the z-axis in the y–z plane due to the fact that the feed point is located in arm filament 1.

Formation of these radiation patterns can be understood from the phase of the current on the loop. At f_{N1} and f_{H1}, the currents on arm filaments 1 and 3 are in-phase (the currents on arm filaments 2 and 4 are not in-phase) and hence single-peak radiation in the z-direction is obtained. At f_{N2} and f_{H2}, the currents on arm filaments 1 and 3 are out-of-phase, and those on arm filaments 2 and 4 are also out-of-phase. As a result, the metaloop forms dual-peak radiation.

The solid line in Fig. 20.4 shows the frequency response of the gain in the direction of maximum radiation intensity (Gain for a balanced mode excitation: Balmex). The gains only at frequencies around f_{N1}, f_{H1}, f_{N2}, and f_{H2} are depicted. It is found that the gain for the single-peak radiation at f_{N1} is smaller than that at f_{H1}, and the gain for the dual-peak radiation at f_{N2} is smaller than that at f_{H2}. This is attributed to the fact that the electrical antenna size (defined by the peripheral length (PL) relative to the free-space wavelength) at frequency f_{N1}, that is, PL/λ_{N1}, is smaller than PL/λ_{H1} at f_{H1}; a similar relationship exists for the electrical antenna sizes at f_{N2} and f_{H2}. Note that the dotted lines show the gain for an unbalanced-mode excitation metaloop (Unbalmex), which will be discussed later in Exercise [3].

The input characteristic in terms of the VSWR (relative to a natural characteristic impedance of $Z_{NTR} = 80\,\Omega$) is shown by the solid lines in Fig. 20.5. It is revealed that the metaloop resonates at frequencies near f_{N2}, f_{N1}, f_{H1}, and f_{H2}: 1λ_g-resonance near f_{N1} and f_{H1} and 2λ_g-resonance near f_{N2} and f_{H2}. Note that the dotted lines represent the answer to Exercise, which will appear later.

Exercise

Investigate the metaloop excited by an unbalanced-mode feed shown in Fig. 20.6. Use the same parameters as those in Table 20.1.

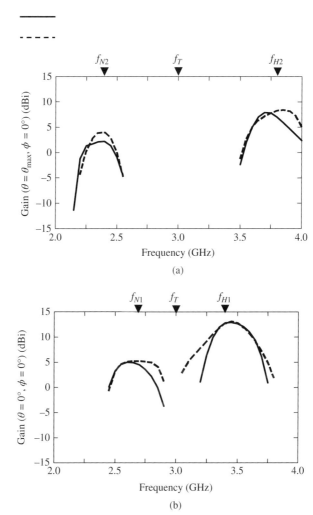

Figure 20.4 Frequency response of the gain. (a) At frequencies around f_{N2} and f_{H2} for dual-peak radiation. (b) At frequencies around f_{N1} and f_{H1} for single-peak radiation.

Answer Figure 20.7 shows the radiation patterns at frequencies near f_{N1}, f_{N2}, f_{H1}, and f_{H2}. The components E_θ and E_ϕ for the unbalanced-mode excitation (UnBalmex) metaloop correspond to E_ϕ and E_θ for the balanced-mode excitation (Balmex) metaloop (see Fig. 20.3), respectively, that is, an exchange in the field components occurs.

The gain for the Unbalmex metaloop is depicted by the dotted lines in Fig. 20.4. It is found that the maximum gain for the Unbalmex is almost the same as that for the Balmex metaloop. The frequency response of the VSWR for the Unbalmex is similar to that for the Balmex, except for the high-frequency resonance around f_{H2}, as shown by the dotted lines in Fig. 20.5. ∎

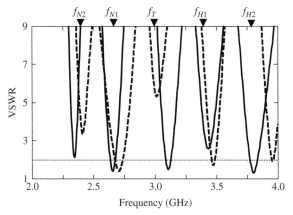

Figure 20.5 Frequency response of the VSWR (relative to $Z_{NTR} = 80\,\Omega$). Note that the dotted lines show the VSWR for an unbalanced-mode excitation metaloop (Unbalmex), which will be discussed later in Exercise E20.2. (Reproduced from Ref. [3] with permission from IEICE.)

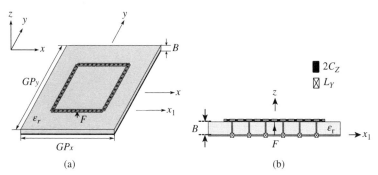

Figure 20.6 Unbalanced-mode excitation metaloop antenna. (a) Perspective view. (b) Side view along line x_1.

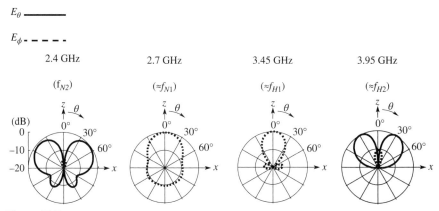

Figure 20.7 Radiation patterns at frequencies near f_{N1}, f_{N2}, f_{H1}, and f_{H2}.

REFERENCES

1. H. Nakano, M. Miura, K. Yoshida, and J. Yamauchi, Metaloop for linearly polarized radiation. In: *Proc. IEEE-APS Topical Conference on Antennas and Propagation in Wireless Communications*, Palm Beach, Aruba, August 2014, pp. 15–17.
2. H. Nakano, M. Miura, K. Yoshida, and J. Yamauchi, LP radiation from a metamaterial-based loop antenna. In: *Proc. Int. Sympo. Antennas Propag.*, ISAP, Kaohsiung, December 2014, pp. 191–192.
3. Y. Kobayashi, J. Yamauchi, and H. Nakano, Linearly polarized metaloop antenna excited by a single line. In: *Proc. IEICE Society Conference*, Tokushima, September 2014, p. B-1-71.

Chapter 21

Circularly Polarized Metaloop Antenna

Chapter 20 presents a metaloop that has a *closed* structure and is excited in balanced and unbalanced modes. It is found that the radiation from the closed metaloop forms linearly polarized (LP) single- and dual-peak radiation, depending on the operating frequency. This chapter discusses a different metamaterial-based loop that has an *open* structure and is excited in unbalanced mode [1–4]. The open metaloop is designed such that it radiates a right-handed circularly polarized (RH CP) wave across a specific frequency range and a left-handed circularly polarized (LH CP) wave across a different frequency range (dual-band counter-CP radiation). For this, one arm end of the open metaloop is connected to the inner conductor of a coaxial feed line and the other end is terminated through a resistive load (R_B) to the ground plane. Note that natural loop antennas fed from a single point [5,6] cannot generate such dual-band counter-CP radiation.

21.1 CONFIGURATION

Figure 21.1 shows an open metaloop. Point F is the arm starting point (feed point) and point T is the arm end shorted to the ground plane through a resistive load R_B. As in the closed metaloop, the antenna arm is composed of small, iterative, conducting strip cells printed on a dielectric substrate. The parameters that differ from those for the closed metaloop in the previous chapter are w, L_{side}, $2C_Z$, and L_Y. These are, respectively, chosen to be 8.8 mm, 50 mm, 1.3 pF, and 1.2 nH for the open metaloop. Note that L_0 and R_B for the open metaloop are chosen to be 20 mm and 50 Ω, respectively.

21.2 COUNTER-CP RADIATION

Figure 21.2 shows the electrical loop peripheral length as a function of frequency, where the physical peripheral length ($= 4L_{side}$) is normalized to the guided

Low-Profile Natural and Metamaterial Antennas: Analysis Methods and Applications, First Edition. Hisamatsu Nakano.

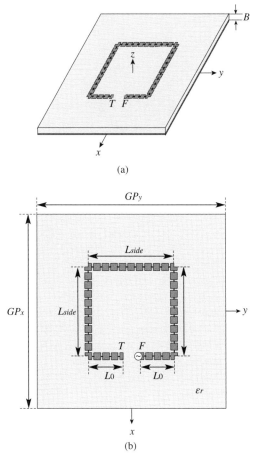

Figure 21.1 Open metamaterial loop antenna. (a) Perspective view. (b) Top view. (Reproduced from Ref. [2] with permission from IEEE.)

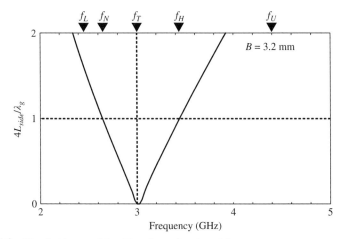

Figure 21.2 Electrical loop peripheral length as a function of frequency.

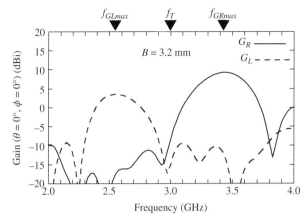

Figure 21.3 Frequency response of the gain.

wavelength of the current flowing on the antenna arm (λ_g). Note that the natural characteristic impedance for the base line is $Z_{NTR} = 50\,\Omega$ and the transition frequency is chosen to be $f_T = 3\,\text{GHz}$. f_N and f_H are referred to as the *nion frequency* and *hion frequency*, respectively, where $4L_{side}/\lambda_g = 1$.

The phase constant β is negative at frequencies below f_T and hence it is expected that LH CP radiation will occur from the metaloop arm (wound in the counterclockwise direction from starting point F) across the low-frequency band. Conversely, the phase constant β is positive at frequencies above f_T and hence it is expected that RH CP radiation will occur across the high-frequency band. Such a characteristic is not obtained with natural loop antennas.

The aforementioned dual-band counter CP radiation is confirmed in Fig. 21.3, where G_L and G_R are the gain for an LH CP wave and an RH CP wave, respectively. As expected, G_L dominates in the low-frequency band and G_R dominates in the high-frequency band. The gain bandwidth for LH CP radiation is approximately 12%, with a maximum gain of $G_L = 3.5$ dBi at frequency f_{GLmax}; the gain bandwidth for RH CP radiation is approximately 10%, with a maximum gain of $G_R = 9.3$ dBi at frequency f_{GRmax} (both are for a 3 dB reduced gain criterion).

Figure 21.4 shows the radiation patterns when G_L and G_R exhibit their maximum value at frequencies f_{GLmax} and f_{GRmax}, respectively, where the radiation is decomposed into an RH CP field component (E_R) and an LH CP radiation field component (E_L). These radiation patterns, together with Fig. 21.3, reconfirm the presence of dual-band counter-CP radiation. Note that the current on the arm travels in a leaky wave fashion, and hence a perfectly symmetric radiation pattern with respect to the z-axis cannot be obtained, as observed. However, the deterioration in the symmetry is not remarkable. Also note that the ratio of the guided wavelength to the free-space wavelength (λ_g/λ) at frequency f_{GLmax} is smaller than that at frequency f_{GRmax}, leading to a difference in the beam widths.

The frequency response of the axial ratio at frequencies around f_{GLmax} and f_{GRmax} is shown in Fig. 21.5. This axial ratio is the value in the z-direction. Within

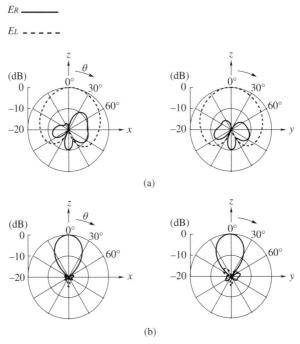

Figure 21.4   Radiation patterns. (a) At $f_{GLmax} = 2.550\,\text{GHz}$. (b) At $f_{GRmax} = 3.425\,\text{GHz}$.

the bandwidth for a 3 dB axial ratio criterion, the VSWR (relative to $50\,\Omega$) is small, as shown in Fig. 21.6. The power input to the open metaloop, P_{in}, is transformed into radiation, P_{rad}, and any remaining power, $P_{in} - P_{rad}$, is absorbed by a resistive load near the arm end, T. Figure 21.7 shows the frequency response of the radiation efficiency (defined by $\eta_{rad} = P_{rad}/P_{in}$). If higher radiation efficiency is required, it is recommended that the antenna height B be increased.

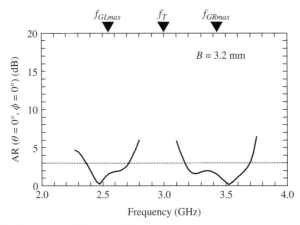

Figure 21.5   Axial ratio around f_{GLmax} and f_{GRmax}.

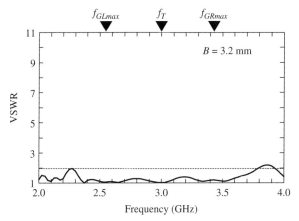

Figure 21.6   VSWR around f_{GLmax} and f_{GRmax}.

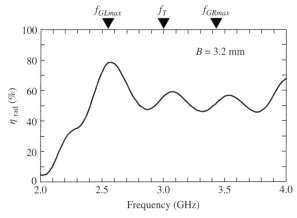

Figure 21.7   Radiation efficiency.

Exercise

The maximum gain for G_L (G_{Lmax}) is smaller than the maximum gain for G_R (G_{Rmax}), as shown in Fig. 21.3. Make the difference between the maximum gains G_{Lmax} and G_{Rmax} be less than 1 dB.

Answer   We use a Fabry–Pérot resonator concept for the solution. Figure 21.8 shows the metaloop discussed above, together with a dielectric slab of thickness B_s and relative permittivity ε_r. The distance between the metaspiral and the dielectric slab d is chosen to be half the wavelength at $f_{GLmax} = 2.55$ GHz. When B_s is appropriately chosen, a difference in the maximum gains G_{Rmax} and G_{Lmax} is found to be less than 1 dB, as shown in Fig. 21.9 [7]   ■

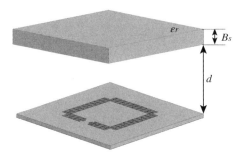

Figure 21.8 Metaloop with a dielectric slab.

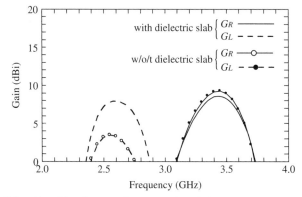

Figure 21.9 Gain with a slab, where $B_s = 21.5$ mm and $\varepsilon_r = 2.6$.

REFERENCES

1. H. Nakano, K. Yoshida, and J. Yamauchi, Counter circularly-polarized loop antenna. In: *Proc. IEEE APS Int. Sympo. Antennas Propag.*, Orlando, FL, July 2013, pp. 152–153.
2. H. Nakano, K. Yoshida, and J. Yamauchi, Radiation characteristics of a metaloop antenna. *IEEE AWPL*, vol. 12, pp. 861–863, 2013.
3. H. Nakano, K. Yoshida, and J. Yamauchi, Dual-band metaloop antenna. *IEEE International Workshop Antenna Technology (iWAT)*, Sydney, Australia, March 2014, pp. 262–265.
4. H. Nakano, K. Yoshida, and J. Yamauchi, Left-handed loop antenna. In: *Proc. IEEE APS Int. Sympo. Antennas Propag.*, Memphis, TN, July 2014, pp. 1550–1551.
5. K. Hirose, T. Haraga, and H. Nakano, A circularly polarized loop antenna without perturbation segments. In: *Proc. IEEE APS Int. Sympo. Antennas Propag.*, July 2010, pp. 1–4.
6. R. Li, V. Fusco, and H. Nakano, Circularly polarized open-loop antenna. *IEEE Trans. Antennas Propag.*, vol. 51, no. 9, pp. 2475–2477, 2003.
7. K. Yoshida, J. Yamauchi, and H. Nakano, Metaloop antenna with a dielectric slab. In: *Proc. IEICE General Conference*, Shiga, March 2015, p. B-1-85.

Chapter 22

Metaspiral Antenna

$\mathbf{A}$ natural spiral antenna has a circumference of greater than one wavelength at the operating frequency [1]. The currents on the spiral arms flow in a traveling wave fashion with a positive phase constant ($\beta > 0$). Therefore, the polarization of the radiation is uniquely determined by the winding direction of the spiral arms, that is, the natural spiral radiates either a right-handed circularly polarized (RH CP) wave or a left-handed circularly polarized (LH CP) wave in a specific direction. It is not possible for the natural spiral to simultaneously radiate both RH CP and LH CP waves (counter-CP radiation).

The spiral presented in this chapter eliminates the above limitation on the CP radiation by realizing a metamaterial property (LH property) for the antenna arms. This spiral antenna is referred to as a metaspiral antenna (invented in 2011 [2]). The metaspiral is capable of dual-band counter-CP radiation. The effect of the size of the metaspiral on the antenna characteristics, including the gain, radiation pattern, axial ratio, and VSWR, is investigated [2,3]. The frequency separation of the dual band (band separation) is also discussed.

22.1 CIRCULARLY POLARIZED RADIATION

22.1.1 Left-Handed Circularly Polarized Wave

Figure 22.1 shows a natural spiral antenna composed of arms A and B. These arms are wound symmetrically with respect to the origin of the rectangular coordinate system. The innermost points of arms A and B are denoted as F_A and F_B and the outermost points are denoted as T_A and T_B, respectively.

Figure 22.1a shows a situation at frequency f_{LH}, where the current at point T_A on arm A and the current at point T_B on arm B are assumed to have a 180° phase difference and travel from these outermost points toward innermost points F_A and F_B. With the assumption of a 180° phase difference, arrows I_{A-U} and I_{B-V} are in the same direction, where I_{A-U} depicts the current element at U_A (near point U) on arm A and I_{B-V} depicts the current element at V_B (near point V) on arm B (note that points V and U are symmetric with respect to the spiral center).

Low-Profile Natural and Metamaterial Antennas: Analysis Methods and Applications, First Edition.
Hisamatsu Nakano.

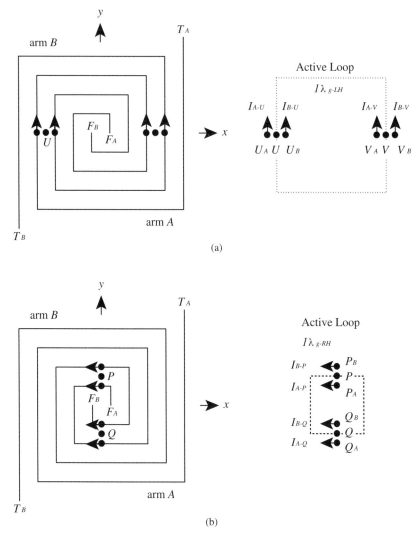

Figure 22.1 Two-arm spiral antenna. (a) Left-handed circularly polarized radiation. (b) Right-handed circularly polarized radiation. (Reproduced from Ref. [2] with permission from IEEE.)

As the A-arm current travels toward the innermost point F_A, it experiences a phase change. When the path length from point U_A to point V_A (near point V) is half the guided wavelength ($\lambda_{g\text{-}LH}/2$ at frequency f_{LH}), the A-arm current experiences a $180°$ phase shift. This is depicted by arrow $I_{A\text{-}V}$ (the A-arm current element near point V), whose direction is the same as the direction for the neighboring B-arm current element arrow $I_{B\text{-}V}$. Thus, a pair of in-phase currents, called a *current band*, is produced near point V.

Similarly, the B-arm current, traveling toward the innermost point F_B, experiences a phase shift of $180°$ along the path from point V_B to point U_B (near point U), because the path length is $\lambda_{g\text{-}LH}/2$ from point V_B to point U_B. The direction of this current element, denoted by arrow $I_{B\text{-}U}$, and the direction of the neighboring current element $I_{A\text{-}U}$ are in the same direction, producing a current band near point U.

As time t changes, these two current bands rotate in the clockwise direction along a square-loop region whose circumference is one guided wavelength ($1\lambda_{g\text{-}LH}$) on the spiral plane. As a result, the spiral radiates an LH CP wave in the positive z-direction.

The currents in this situation experience a delay in phase as they travel from points T_A and T_B toward points F_A and F_B, respectively; this is equivalent to currents with a negative phase constant ($\beta < 0$) flowing from F_A and F_B toward T_A and T_B, respectively. In other words, the spiral with feed points F_A and F_B radiates an LH CP wave if the phase constant of the current is negative.

22.1.2 Right-Handed Circularly Polarized Wave

Another situation is considered, where the direction of the currents flowing at frequency f_{RH} ($>f_{LH}$) is opposite to that in the previous situation, that is, the currents at innermost points F_A and F_B are assumed to have a 180° phase difference and travel from these points toward outermost points T_A and T_B, respectively. This situation is shown in Fig. 22.1b. Due to the 180° phase difference, the current element on arm A near point P, denoted by arrow $I_{A\text{-}P}$, and the current element on arm B near point Q, denoted by arrow $I_{B\text{-}Q}$, are in the same direction, where points P and Q are symmetric with respect to the spiral center. When the path length from point P_A (near point P) to point Q_A (near point Q) is half the guided wavelength ($\lambda_{g\text{-}RH}/2$ at frequency f_{RH}), the A-arm current experiences a phase shift of 180°. As a result, the A-arm current element at Q_A, denoted by arrow $I_{A\text{-}Q}$, and the neighboring B-arm current element arrow $I_{B\text{-}Q}$ are in the same direction. Thus, a current band is produced near point Q.

Note that, due to the same mechanism, another current band is produced near point P; the B-arm current element at point P_B (near point P), denoted by arrow $I_{B\text{-}P}$, and the neighboring A-arm current element arrow $I_{A\text{-}P}$ are in the same direction.

The abovementioned two current bands rotating with time t along a $1\lambda_{g\text{-}RH}$ square-loop region generate RH CP radiation in the positive z-direction. Note that this RH CP radiation is generated by choosing innermost points F_A and F_B as the antenna feed points, making the phase constant of the currents positive ($\beta > 0$) relative to the winding direction of the arms, starting from the innermost points and ending at the outermost points.

22.1.3 Configuration of a Metaspiral Antenna

The previous section concludes that a spiral antenna fed from points F_A and F_B radiates an RH CP wave at frequency f_{RH}. If the phase constant is negative at frequency f_{LH}, then the same spiral (again fed from points F_A and F_B, not T_A and T_B) radiates an LH CP wave at frequency f_{LH}. Based on this result, two arms with a metamaterial property are realized, as shown in Fig. 22.2, where a square-ground

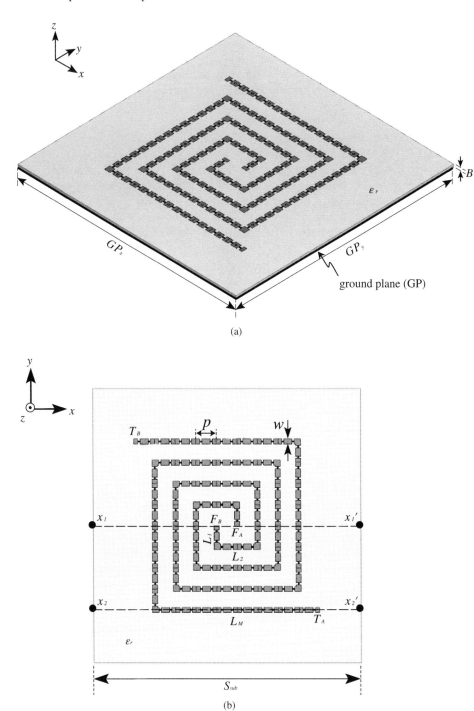

(a)

(b)

Figure 22.2   Two-arm metaspiral antenna. (a) Perspective view. (b) Top view. (c) Side view.

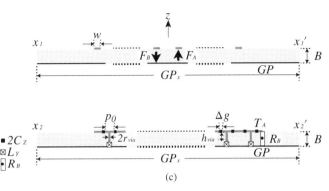

Figure 22.2   (Continued)

plane (side length $GP_x = GP_y$) and a square dielectric substrate (side length $S_{sub} = GP_x = GP_y$, relative permittivity ε_r, and thickness B) are used. Each arm is composed of M straight filaments, whose lengths are labeled $L_1, L_2, \ldots, L_M$, where the mth filament length is defined by $L_m = mL_1$. The innermost points F_A and F_B are used as the feed points (fixed) and the outermost points T_A and T_B are shorted to the ground plane through a resistive impedance, R_B. This two-arm spiral antenna is referred to as a metaspiral.

As will be revealed later, the metaspiral operates with a much smaller antenna height than the corresponding natural antenna, for example, the metaspiral operates with an antenna height of approximately $\lambda/100$ (with λ being the free-space operating wavelength). Note that natural spirals with such a small antenna height do not work as a CP antenna.

22.1.4   Dual-Band Counter-CP Antenna Characteristics

For the following discussion, the parameters shown in Tables 19.1 and 19.2 are used for w, B, ε_r, p, p_0, Δg, r_{via}, and h_{via}, together with $L_1 = 10\,\text{mm}$. The base-line natural characteristic impedance Z_{NTR} is approximately $80\,\Omega$. The transition frequency f_T is set to be 3 GHz by adjusting the value of a series capacitance $2C_Z$ and parallel inductance L_Y.

A length of $4L_M$ is referred to as the antenna peripheral length (PL). For CP radiation, the antenna peripheral length is chosen to be large enough to support the current bands on the spiral plane, that is, $4L_M > 1\lambda_g$, where λ_g is the guided wavelength at frequency f. Figure 22.3 shows the antenna peripheral length relative to the guided wavelength, λ_g. As the number of filaments M is increased, the nion frequency f_N and the hion frequency f_H move toward the transition frequency f_T. Consequently, it is expected that the maximum-gain frequency for an LH CP wave, f_{GLmax}, and the maximum-gain frequency for an RH CP wave, f_{GRmax}, will move toward f_T. This is confirmed by Fig. 22.4, where G_L and G_R are the gains for LH

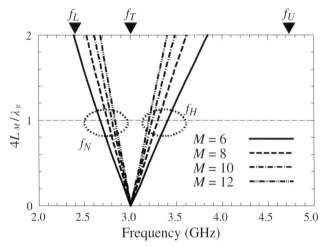

Figure 22.3   Antenna peripheral length (PL = 4L$_M$) as a function of frequency. (Reproduced from Ref. [2] with permission from IEEE.)

CP and RH CP waves, respectively. In addition, it is found that the maximum gain for G_L and G_R increases as the number of filaments M is increased.

Figure 22.5 shows the radiation patterns at the maximum-gain frequencies f_{GLmax} and f_{GRmax}. It is found that the half-power beam width (HPBW) of the radiation pattern for the LH CP wave component is wider than that for the RH CP wave component, due to a difference in the antenna peripheral length relative to the free-space operating wavelength. The HPBW as a function of the number of filaments M is shown in Fig. 22.6. It is revealed that, as M is increased, the HPBW decreases, which is associated with an increase in the gain, as shown in Fig. 22.5.

The resistive termination for the antenna arms stabilizes the input impedance (and hence the VSWR). The solid line in Fig. 22.7 shows a representative VSWR relative to $R_B = 80\,\Omega$ at frequencies around the maximum-gain frequencies f_{GLmax} and f_{GRmax}. As desired, the VSWR is found to be small. Note that a dotted line showing the VSWR relative to $50\,\Omega$ is added to this figure as additional information.

22.2   LINEARLY POLARIZED RADIATION

The dual-band counter-CP radiation at frequencies around the maximum-gain frequencies f_{GLmax} and f_{GRmax} is discussed in the previous section. This section reveals linearly polarized (LP) radiation [4], which becomes stronger as the phase constant β of the currents approaches zero. It is worthwhile to determine the direction of the linear polarization, that is, the azimuth angle ϕ_p from the x-axis, referred to as the polarization azimuth angle (PAA). Figure 22.8 shows the PAA as a function of

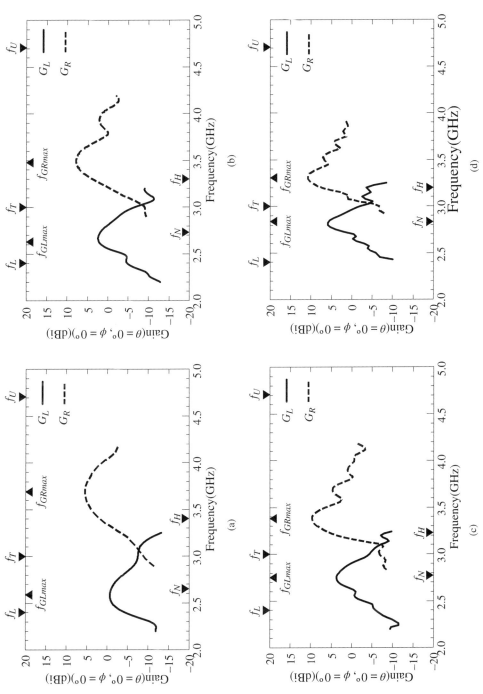

Figure 22.4 Frequency response of the gain. (a) Number of filaments $M = 6$ (antenna peripheral length $= 4L_6$). (b) $M = 8$ (antenna peripheral length $= 4L_8$). (c) $M = 10$ (antenna peripheral length $= 4L_{10}$). (d) $M = 12$(antenna peripheral length $= 4L_{12}$).

267

E_R ———

E_L - - - - -

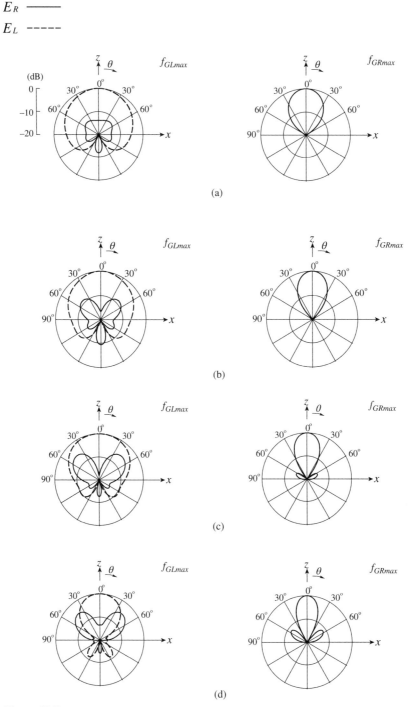

Figure 22.5   Radiation patterns at the maximum-gain frequencies f_{GLmax} and f_{GRmax}. (a) Number of filaments $M = 6$. (b) $M = 8$. (c) $M = 10$. (d) $M = 12$.

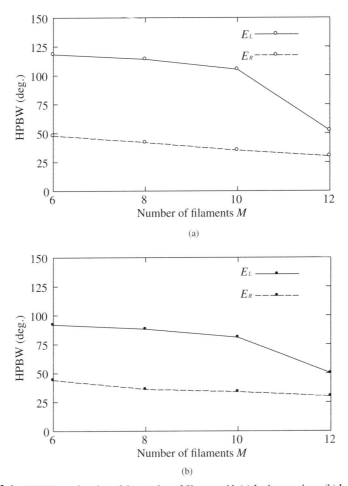

Figure 22.6 HPBW as a function of the number of filaments *M*. (a) In the *x–z* plane. (b) In the *y–z* plane.

the number of cells, and the number of arm filaments *M*. Note that the PAA is calculated at a frequency where the gain G_L equals G_R. The PAA is found to change linearly with *M*. This means that the PAA can be controlled by appropriately choosing the number of arm filaments *M*.

Figure 22.9 shows a representative LP radiation pattern, where E_θ and E_ϕ are the θ and ϕ components of the radiated electric field, respectively, in the spherical coordinate system. Note that the HPBW of the E_θ component becomes narrower as the number of arm filaments M is increased, resulting in an increase in the gain for an LP wave.

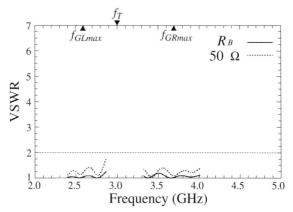

Figure 22.7   Frequency response of the VSWR for $M=6$. The solid and dotted lines show the VSWRs relative to $R_B = 80$ and $50\,\Omega$, respectively. (Reproduced from Ref. [2] with permission from IEEE.)

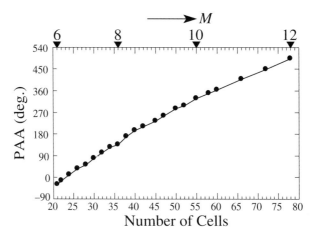

Figure 22.8   Polarization azimuth angle (PAA) as a function of the number of cells (and arm filaments M).

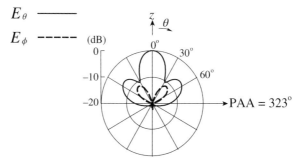

Figure 22.9   LP radiation pattern for $M=10$.

REFERENCES

1. C. Balanis, *Modern Antenna Handbook*, Chap. 6, Wiley, 2008.
2. H. Nakano, J. Miyake, M. Oyama, and J. Yamauchi, Metamaterial spiral antenna. *IEEE Antennas Wireless Propag. Lett.*, vol. 10, pp. 1555–1558, 2011.
3. H. Nakano, J. Miyake, T. Sakurada, and J. Yamauchi, Dual-band counter circularly polarized radiation from a single-arm metamaterial-based spiral antenna. *IEEE Trans. Antennas Propagat.*, vol. 61, no. 6, pp. 2938–2947, 2013.
4. H. Nakano, H. Kataoka, T. Shimizu, and J. Yamauchi, Two-arm metaspiral antenna. In: *Proc. EuCAP*, The Hague, April 2014, pp. 788–790.

Chapter 23

Metahelical Antennas

The current distribution of a natural end-fire helical antenna is presented in Chapter 17. One finding is that there are two distinct current regions when the number of helical turns is large (resulting in a long helical antenna): the C-region and the S-region. The C-region generates a circularly polarized (CP) wave, and the S-region acts as a director for the CP wave radiated from the C-region. This means that a natural *low-profile* antenna composed of only the C-region (in the absence of the S-region) can be an end-fire CP element, whose polarization is either left-handed (LH) or right-handed (RH), depending on the arm winding sense.

This chapter presents a low-profile end-fire helical antenna that differs from the natural end-fire helical antenna in that it acts as a dual-band counter-CP radiation element; it radiates an LH CP wave across a specific frequency band and an RH CP wave across a different frequency band [1–4], similar to the metaspiral in Chapter 22. This dual-band counter-CP helical antenna is referred to as *a metahelical antenna*.

23.1 ROUND METAHELICAL ANTENNA

23.1.1 Configuration

Figure 23.1 shows the configuration of a metahelical antenna, where the antenna circumference is defined by $L_c = 2\pi r_{arm}$, with r_{arm} being the radial distance from the z-axis to the helical arm. Unlike natural helical antennas consisting of continuous arms, the metahelical antenna has a discontinuous arm, which is a metaline wound on a cylindrical dielectric layer (relative permittivity ε_r and thickness B) backed by a curved conducting plane (GP′) parallel to the cylindrical dielectric layer. This conducting plane is electrically connected to the bottom ground plane (GP).

The configuration parameters used here are shown in Table 23.1, where the metaline is specified by the same notation as that for the metaline shown in Fig. 19.3. The base-line natural characteristic impedance and the transition frequency are set to be $Z_{NTR} = 80\,\Omega$ and $f_T = 3\,\text{GHz}$, respectively.

Low-Profile Natural and Metamaterial Antennas: Analysis Methods and Applications, First Edition. Hisamatsu Nakano.
© 2016 The Institute of Electrical and Electronics Engineers, Inc. Published 2016 by John Wiley & Sons, Inc.

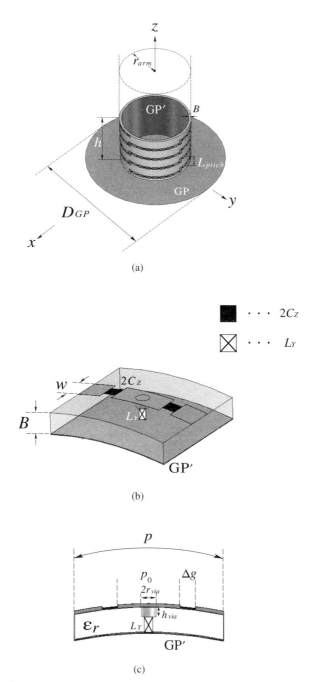

Figure 23.1   Metahelical antenna. (a) Perspective view. (b) Unit cell above a curved conducting plane GP′. (c) View of the unit cell from the z-direction.

Table 23.1 Configuration Parameters

Symbol	Value	Symbol	Value (mm)	Symbol	Value
w	4.4 mm	Δg	1.0	D_{GP}	∞
B	1.6 mm	p	10.0	r_{arm}	24.2 mm
ε_r	2.6	r_{via}	0.5	L_{pitch}	8.0 mm
p_0	4.0 mm	h_{via}	0.6	h	Varied

23.1.2 Antenna Characteristics

Based on the behavior of the current along the metahelical arm, it is expected that the maximum gain for an LH CP wave, G_{Lmax}, will appear at a frequency below f_T, where the phase constant is negative ($\beta < 0$) and the antenna circumference normalized to the guided wavelength, L_c/λ_g, corresponds to approximately one; in addition, the maximum gain for an RH CP wave, G_{Rmax}, will appear at a different frequency above f_T, where the phase constant is positive ($\beta > 0$) and the normalized antenna circumference L_c/λ_g again corresponds to approximately one.

Figure 23.2 shows one example of the frequency response of the gain, where the number of helical turns N is fixed to be four. Note that G_L and G_R in this figure denote the gain for an LH CP wave and the gain for an RH CP wave, respectively. As expected, the maximum gain, G_{Lmax} and G_{Rmax}, appears in the bore sight direction (z-direction) at a frequency near the Nion frequency f_N and at a frequency near the Hion frequency f_H. These two frequencies are denoted as f_{GLmax} and f_{GRmax}, respectively. Note that the antenna height is small: 0.29 wavelength at f_{GLmax}.

Figure 23.3 shows the maximum gain, G_{Lmax} and G_{Rmax}, as a function of the number of helical turns N. G_{Lmax} is less sensitive to the number of helical turns for $N > 3$, while G_{Rmax} increases as N is increased. This means that the array effect due

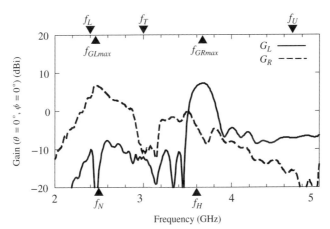

Figure 23.2 Frequency response of the gain, where the number of helical turns N is four. The parameters in Table 23.1 are used.

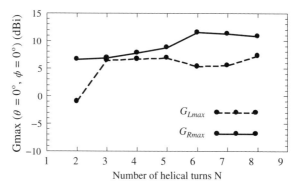

Figure 23.3   Maximum gain, G_{Lmax} and G_{Rmax}, as a function of the number of helical turns N.

to the helical turns for gain G_{Lmax} is weaker than that for G_{Rmax}, resulting from the different electrical pitch length at frequencies f_{GLmax} and f_{GRmax} ($L_{pitch}/\lambda_{GLmax} \neq L_{pitch}/\lambda_{GRmax}$, where λ_{GLmax} and λ_{GRmax} are the free-space wavelengths at f_{GLmax} and f_{GRmax}, respectively). Note that $L_{pitch}/\lambda_{GLmax}$ for the LH CP radiation is smaller than $L_{pitch}/\lambda_{GRmax}$ for the RH CP radiation.

Decomposing the radiation field into an LH CP electric field component E_L and an RH CP electric field component E_R helps in understanding the metahelical antenna. Figure 23.4a illustrates the radiation pattern at the maximum-gain frequency for an LH CP wave, f_{GLmax}, and at the maximum-gain frequency for an RH CP wave, f_{GRmax}, where the number of helical turns is four ($N = 4$). Similarly, Fig. 23.4b illustrates the radiation patterns for $N = 6$. The asymmetry in the radiation pattern is due to the antenna structure, which is not symmetric with respect to the z-axis. However, as the number of helical turns is further increased from $N = 6$, the copolarization component (E_L at f_{GLmax} and E_R at f_{GRmax}) becomes relatively symmetric with respect to the z-axis. Note that these copolarization component patterns have different half-power beam widths, due to the dispersive nature of the current along the antenna arm. Also note that, as the number of helical turns is increased, the cross-polarization component becomes smaller.

The end of the metahelical arm is terminated to the curved conducting plane through a resistive load, which reduces reflected waves and improves the input characteristic (VSWR). Figure 23.5 shows the VSWR at frequencies around the maximum-gain frequencies f_{GLmax} and f_{GRmax}. The VSWR is found to be small within these two frequency regions.

23.2   RECTANGULAR METAHELICAL ANTENNA

Section 23.1 presents a round metahelical antenna whose arm is wound on a cylindrical dielectric layer. Figure 23.6 shows a rectangular metahelical antenna as a counterpart to the round metahelical. The helical arm is made of a bent metaline [2–4]. The notation for the configuration parameters of this metaline is the same as that for the metaline shown in Fig. 19.3.

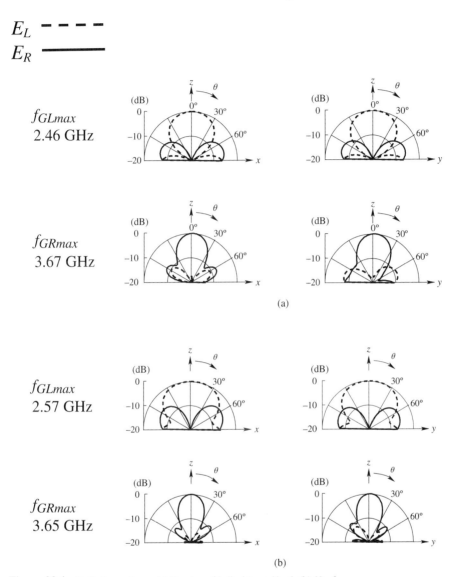

Figure 23.4 Radiation patterns. (a) Number of helical turns $N=4$. (b) $N=6$.

Figure 23.7 shows the frequency response of the gains G_L and G_R, where the configuration parameters used are summarized in Table 23.2. Note that the transition frequency and the base-line natural characteristic impedance are chosen to be $f_T=3$ GHz and $Z_{NTR}=80\,\Omega$, respectively. It is found that the bandwidth for a 3 dB reduced gain criterion (shaded region) becomes narrower as the number of helical turns N is increased.

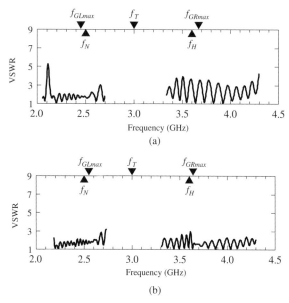

Figure 23.5   Frequency response of the VSWR. (a) Number of helical turns $N = 4$. (b) $N = 6$.

Exercise

Reveal the radiation pattern and VSWR when the number of rectangular metahelical turns is $N = 2$, 3, and 4. Use Table 23.2.

Answer   Figure 23.8 shows the radiation pattern. The side lobes at f_{GLmax} become smaller as the N is increased. The VSWR across the gain bandwidth (shaded region in Fig. 23.7) is reasonably small, as shown in Fig. 23.9.   ∎

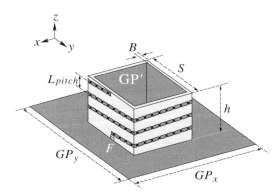

Figure 23.6   Rectangular metahelical antenna. (Reproduced from Ref. [2] with permission from IEEE.)

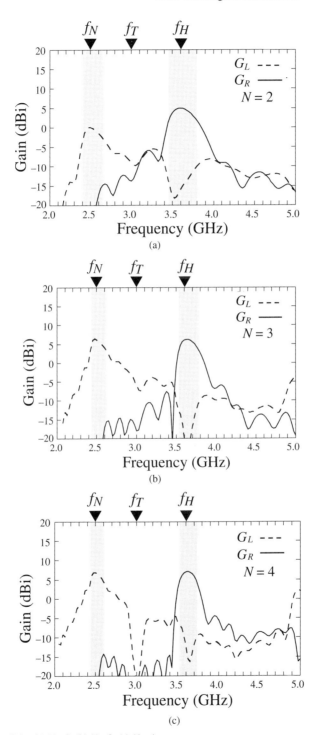

Figure 23.7 Gain. (a) $N=2$. (b) $N=3$. (c) $N=4$.

Table 23.2   Configuration Parameters for the Rectangular Metahelical Antenna

Symbol	Value	Symbol	Value	Symbol	Value
w	2.0 mm	p	10 mm	GP_y	∞
B	1.6 mm	r_{via}	0.5 mm	S	39.9 mm
ε_r	2.6	h_{via}	1.0 mm	h	27.6 mm
p_0	4.0 mm	GP_x	∞	L_{pitch}	8.0 mm
Δg	1.0 mm				

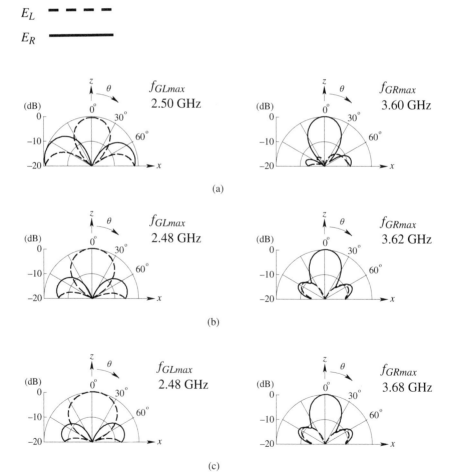

Figure 23.8   Radiation patterns. (a) $N=2$. (b) $N=3$. (c) $N=4$.

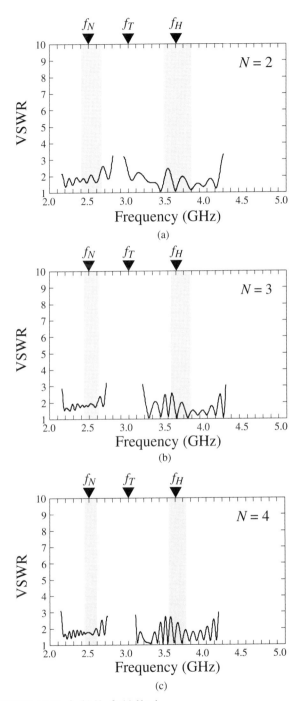

Figure 23.9   VSWR. (a) $N = 2$. (b) $N = 3$. (c) $N = 4$.

REFERENCES

1. H. Nakano, M. Tanaka, and J. Yamauchi, Radiation from metahelical antenna. In: *Proc. Int. Sympo. Antennas Propag. (ISAP)*, vol. 2, Nanjing, China, October 2013, pp. 1206–1207.
2. H. Nakano, M. Tanaka, and J. Yamauchi, Metahelical antenna using a left-handed property. In: *Proc. IEEE-APS Topical Conference on Antennas and propagation in Wireless Communications (APWC)*, Torino, 2013, pp. 74–77.
3. K. Monma, M. Tanaka, J. Yamauchi, and H. Nakano, Metahelical antenna composed of parallelo-gram-formed cells on the radiation characteristics. In: *Proc. IEICE Society Conference*, Tokushima, September 2014, p. BCS-1-15.
4. K. Monma, M. Tanaka, J. Yamauchi, and H. Nakano, Effects of the number of helical turns on the radiation characteristics of a metahelical antenna composed of parallelogram-formed cells. In: *Proc. IEICE General Conference*, Shiga, 2015, p. B-1-109.

Index

Low-Profile Natural and Metamaterial Antennas: Analysis Methods and Applications, First Edition.
Hisamatsu Nakano.
© 2016 The Institute of Electrical and Electronics Engineers, Inc. Published 2016 by John Wiley & Sons, Inc.

IEEE PRESS SERIES ON ELECTROMAGNETIC WAVE THEORY

Andreas C. Cangellaris, *Series Editor*
University of Illinois, Urbana-Champaign, Illinois

Multigrid Finite Element Methods for Electromagnetic Field Modeling
Yu Zhu, Andreas C. Cangellaris

Electromagnetic Theory
Julius Adams Stratton

Electromagnetic Fields, Second Edition
Jean G. Van Bladel

Electromagnetic Fields in Cavities: Deterministic and Statistical Theories
David A. Hill

Discontinuities in the Electromagnetic Field
M. Mithat Idemen

Understanding Geometric Algebra for Electromagnetic Theory
John W. Arthur

The Power and Beauty of Electromagnetic Theory
Frederic R. Morgenthaler

Modern Lens Antennas for Communications Engineering
John Thornton, Kao-Cheng

Electromagnetic Modeling and Simulation
Levent Sevgi

Multiforms, Dyadics, and Electromagnetic Media
Ismo V. Lindell

Low-Profile Natural and Metamaterial Antennas: Analysis Methods and Applications
Hisamatsu Nakano